Contemporary Chemistry

A Practical Approach

LEONARD SALAND

J. Weston Walch, Publisher

Portland, Maine

Dedication

To my wife, Dorothy, whose encouragement and help brought this book from an idea to a reality, and to my children, Joel and Erika.

Leonard Saland is chairman of the Physical Science Department at Louis Brandeis High School in New York.

1   2   3   4   5   6   7   8   9   10

ISBN 0-8251-1799-2

Copyright © 1986, 1991
J. Weston Walch, Publisher
P.O. Box 658 • Portland, Maine 04104-0658

Printed in the United States of America

Contents

Contemporary Chemistry: A Practical Approach

Preface   *xi*

Chapter 1   Why Do We Study Chemistry?   **1**

Introduction   **3**

1-1   How Do Scientists Solve Problems?   **3**
1-2   How Can I Work Safely in the Lab?   **6**
1-3   What Are the Safety Rules for the Chemistry Lab?   **6**
1-4   How Should I Write My Lab Report?   **7**
1-5   How Did Different Systems of Measurement Come About?   **10**
1-6   Why Is the Metric System Used All Over the World?   **10**
1-7   How Is Temperature Measured?   **13**
Metric Math Problems   **16**
Written Exercises   **17**
Career Information   **23**

Chapter 2   How Does Matter Change?   **25**

Introduction   **27**

2-1   What Is Matter?   **27**
2-2   How Does Matter Show Itself to Us?   **29**
2-3   What Kind of Changes Can Matter Undergo?   **29**
2-4   What Is Matter Made Of?   **31**
2-5   How Is Energy Involved with Molecules?   **32**
(The Kinetic Molecular Theory)
2-6   How Does Matter Change Its State?   **32**
Written Exercises   **37**

Chapter 3   How Is Matter Constructed?   **41**

Introduction   **43**

3-1   What Are Substances Made Of?   **43**
3-2   What Is the Dalton Atomic Theory?   **44**
3-3   What Are Atoms Made Of?   **44**
3-4   How Are Atoms Structured?   **45**
3-5   How Is the Atomic Number Important?   **46**
3-6   How Is the Nucleus Important?   **46**
3-7   How Do the Electrons Orbit the Nucleus?   **47**
3-8   How Does the Periodic Table Reveal a Picture of the Atom?   **48**
3-9   How Can the Properties of the Elements Be Predicted by Trends
       in the Periodic Table?   **49**
3-10   How Do the Properties of the Elements Change in the Same Period?   **51**
       Written Exercises   **54**

Chapter 4   How Do Atoms Combine?   **61**

Part I

Introduction   **63**

4-1   How Are Molecules of Elements and Compounds Different?   **63**
4-2   How Are Atoms Held Together?   **64**
4-3   How Do the Atoms of Elements Share Electrons?   **66**
4-4   How Do Different Atoms Share Electrons?   **68**
4-5   How Do Atoms of Metals Combine?   **69**
4-6   How Does Metallic Bonding Explain Metallic Properties?   **70**
4-7   What Is a Formula?   **71**
4-8   How Do Atoms Combine?   **72**
4-9   What Is a Radical?   **73**
4-10   How Can Our Valence Table Be Used to Determine Chemical
       Formulas?   **74**
4-11   How Can We Write Formulas that Contain Radicals?   **75**
       Written Exercises   **78**

Part II

4-12   How Are Formula Weights Calculated?   **85**
4-13   What Is the Law of Definite Proportions?   **86**
4-14   How Can We Determine the Empirical Formula?   **88**
4-15   How Can We Count Molecules?   **90**
4-16   How Do Scientists Write Very Large and Very Small Numbers?   **91**

Chapter 5   Of What Are Solutions Made?   93

Part I

Introduction   95

5-1   How Is Seawater Different from Fresh Water?   95
5-2   What Is a Solution?   95
5-3   How Can the Solvent Be Separated from the Solute in a Solution?   98
5-4   What Other Kinds of Solutions Are There?   99
5-5   What Are Polar Molecules?   101
5-6   How Do Salt and Sugar Dissolve in Water?   102
5-7   Why Are Oils Insoluble in Water?   103
5-8   How Does Salt Affect Ice?   103
5-9   How Do Dissolved Solutes Affect the Boiling Point of Water?   104
5-10   How Much Solute Is Enough?   104
5-11   How Much Is Too Much?   106
5-12   How Can We Speed the Rate of Solution?   107
5-13   How Can We Control the Amount of Solute that Can Dissolve?   107
5-14   How Can We Grow Crystals as a Hobby?   109
       Written Exercises   112
       Graph Interpretation   119

Part II

5-15   Determining the Concentration of Like Solutions   122
5-16   Comparing the Concentration of Different Solutions   123
       Molarity Computation Problems   124

Chapter 6   What Are Acids, Bases, and Salts?   127

Part I

Introduction   129

6-1   What Is a Strong Acid?   130
6-2   What Is a Base?   132
6-3   How Can Acids and Bases Be Detected?   132
6-4   How Is Acidity Measured?   133
6-5   How Do Acids and Bases React with Each Other?   134
6-6   Are All Salts Neutral to Litmus?   135
       Written Exercises   138

Part II

6-7   How Is pH Related to Molarity?   142
       Written Exercises   145

Chapter 7 Energy Forms and Nuclear Chemistry: World Problem or Solution? 147

Part I Energy Forms

Introduction 149

7-1 Where Does Our Energy Come From? 149
7-2 How Does the Sun Get Its Energy? 149
7-3 Why Can't We Depend on Fossil Fuels to Meet Our Energy Needs? 151
7-4 Can Solar Energy Solve the Energy Problem? 152
7-5 Can Other Sources of Energy Solve Our Energy Problems? 155

Part II Nuclear Chemistry

7-6 How Was Nuclear Radiation Discovered? 156
7-7 What Is Radioactivity? 156
7-8 How Does Radioactive Decay Change Elements? 158
7-9 How Is Radioactivity Detected? 159
7-10 How Does Radiation Affect Living Things? 159
7-11 How Are Radioisotopes Used? 160
7-12 How Is the Rate of Radioactive Decay Measured? 161
7-13 What Is Nuclear Fission? 161
7-14 How Can a Nuclear Reactor Generate Electricity? 163
7-15 Can Nuclear Energy Solve Our Energy Problems? 164
Written Exercises 167
Career Information 175

Chapter 8 What Is the Chemistry of Construction Materials? 177

Introduction 179

8-1 What Are Construction Materials Made Of? 182
8-2 How Is Cement Made? 184
8-3 How Are Ceramics Made? 186
8-4 How Is Glass Made? 188
8-5 How Does Mortar Hold Bricks Together? 189
8-6 How Is Plaster Made? 190
Written Exercises 192
Career Information 197

Chapter 9   How Are Metals Obtained?   199

Part I

Introduction   201

9-1   Where Are Metals Found?   202
9-2   Why Can't All Metals Be Found Free in Nature?   202
9-3   How Are Metals Extracted from Their Ores?   206
9-4   How Is Iron Extracted from the Earth?   206
9-5   Why Is Steel Preferable to Iron?   209
9-6   How Is Steel Made?   211
9-7   How Is High-Carbon Steel Heat-Treated?   213
9-8   How Is Low-Carbon Steel Heat-Treated?   214
9-9   How Can the Rusting of Steel Be Prevented?   214
     Written Exercises   217

Part II

9-10   How Is Lead Extracted from the Earth?   221
9-11   How Is Copper Extracted from the Earth?   222
9-12   How Is Copper Electrolytically Refined?   223
9-13   How Did Aluminum, Our Most Abundant Metal, Become Available?   225
9-14   How Is Aluminum Liberated from the Earth?   226
9-15   What Is the Metal of Opportunity Today?   227
9-16   How Is Titanium Extracted from Its Ores?   227
     Written Exercises   229
     Career Information   232

Chapter 10   Why Do We Study Carbon and Its Compounds?   233

Part I

Introduction   235

10-1   How Do Carbon Atoms Combine?   236
10-2   How Did Organic Chemistry Originate?   236
10-3   What Are the Main Groups of Hydrocarbons?   237
10-4   How Are the Alkanes Different from Other Families?   237
10-5   What Are Isomers?   239
10-6   How Are the Physical Properties of Aliphatic Compounds Affected by Molecular Weight?   239
10-7   How Are the Alkenes Different from Other Families?   241
10-8   How Are the Alkynes Different from Other Families?   242

10-9   How Can We Determine the Formulas for the
          Aliphatic Hydrocarbons?   **243**
       Written Exercises   **245**

Part II

10-10   How Are Alcohols Made?   **250**
10-11   How Are Alcohols Oxidized?   **252**
10-12   How Are Carbohydrates Fermented?   **253**
10-13   What Is an Organic Acid?   **256**
10-14   How Do Acids React with Alcohols?   **257**
        Written Exercises   **260**

Chapter 11   What Is the Chemistry of Food?   **265**

Introduction   **267**

11-1   Where Does Food Come From?   **267**
11-2   What Do We Need from Food?   **268**
11-3   How Does Cooking Change Our Food?   **275**
11-4   How Can Food Be Preserved?   **279**
       Written Exercises   **282**
       Career Information   **288**

Chapter 12   How Are Chemical Creations Changing Life-Styles?   **289**

Introduction   **291**

12-1   What Are Plastics?   **291**
12-2   What Kinds of Polymers Are There?   **292**
12-3   How Are Polymers Formed?   **297**
12-4   What Are Silicones?   **301**
12-5   How Can Synthetic Polymers Be Disposed Of?   **302**
       Written Exercises   **304**

Chapter 13   What Is the Effect of Chemistry on Our Ecology?   **309**

Introduction   **311**

13-1   What Is "Normal" Air?   **313**
13-2   How Does the Air Support Combustion?   **313**

13-3   What Are Some Types of Combustion?   313
13-4   How Can Fires Be Extinguished?   315
13-5   How Are Rains Formed?   315
13-6   How Does Nature Change Our Air?   317
13-7   How Can Carbon Dioxide Pollute Our Air?   317
13-8   How Safe Is Our Water Supply?   325
13-9   How Do Humans Pollute the Land?   330
13-10 How Can Poisonous Chemicals Be Eliminated?   331
       Written Exercises   334
       Career Information   339

Chapter 14   How Do Some Household Chemicals Work?   341

Introduction   343

14-1   How Is Soap Made?   343
14-2   What Kinds of Soaps Are There?   344
14-3   How Does Soap Clean?   345
14-4   Why Are Detergents Used in Place of Soap?   346
14-5   How Do Drain and Oven Cleaners Work?   347
14-6   Why Is Ammonia Preferable as a Cleanser?   348
14-7   What Is Dry Cleaning?   348
14-8   How Are Spots Removed from Fabrics?   348
14-9   How Does Bleaching Take Place?   349
       Written Exercises   352
       Career Information   356

Answer Section   357

Periodic Table of the Elements   381

Alphabetical Table of the Elements   383

Index   385

Preface

Purpose and Organization of This Book

This text is presented as an alternative to the abstract and theoretical chemistry courses offered by many schools today. The topics conform to the preliminary recommendations of the New York City General Chemistry Syllabus Committee. The chemical theory is designed for application to industrial processes and their impact on our society. The theoretical chapters 1, 2, 3, 4, 5, 6, the first half of 10, and sections 3, 4, 5, and 6 of Chapter 13 are recommended as a core area which should be covered in its entirety. The remainder of the text covers optional areas such as energy (Chapter 7), the chemistry of construction materials (Chapter 8), metallurgy (Chapter 9), organic chemistry (Chapter 10), food chemistry (Chapter 11), polymer (plastics) chemistry (Chapter 12), environmental chemistry (Chapter 13), and household chemicals (Chapter 14). The optional areas are meant to be selected by the teacher.

For the Student

This book is written for you. It is designed to give you an understanding of our modern world through chemistry. It offers you opportunities to strengthen your reading, writing, math, and thinking skills through chemistry. In addition, lifelong careers are suggested to you.

For the Teacher

This book is designed to meet the needs of the hard-pressed teacher. The organization of each chapter should suggest a lesson plan. The instructional objectives are stated at the beginning of each chapter and can be objectively evaluated. Chapter titles (goals) and section titles (objectives) are stated in problem form. Ample student exercises are available for use as homework assignments and meaningful classroom drills. Each chapter is designed to assist you in improving your students' reading, writing, math, and problem-solving skills. "Reading Power" and "Find the Facts" exercises appear at the end of each chapter. Writing skills and thought-type problems are presented as "Mind Expanders" exercises. Math skills receive attention in chapters 1, 3, 4, 5, 6, 7, and 10. Interpretation of graphs is included in Chapter 5. Career infusions at the high school diploma and two- and four-year degree levels are introduced at the ends of chapters 1, 7, 8, 9, 11, 13, and 14. Overall, the author has presented a text that "touches all bases" for the teacher as well as the student.

Chapter 1
Why Do We Study Chemistry?

Instructional Objectives

After completing this chapter, you will be able to:

1. Define Celsius, centigrade, chemistry, controlled experiment, experiment, gram, hypothesis, liter, metric system, science, scientific method, and theory.

2. Outline the steps of the scientific method.

3. Achieve 100% on a lab safety test.

4. Write a lab report with a definite format that follows the scientific method.

5. Make measurements in metric units.

6. Recall the metric prefixes and convert from one metric unit into another (length, mass, volume, and temperature).

7. Relate metric units of mass with volume for water.

8. Solve arithmetic problems using metric units.

Chapter 1 Contents

Introduction   **3**

1-1   How Do Scientists Solve Problems?   **3**

1-2   How Can I Work Safely in the Lab?   **6**

1-3   What Are the Safety Rules for the Chemistry Lab?   **6**

1-4   How Should I Write My Lab Report?   **7**

1-5   How Did Different Systems of Measurement Come About?   **10**

1-6   Why Is the Metric System Used All Over the World?   **10**

1-7   How Is Temperature Measured?   **13**

Metric Math Problems   **16**

Written Exercises   **17**

Career Information   **23**

Chapter 1
Why Do We Study Chemistry?

Introduction

The scientist observes and tries to understand nature. The biologist studies living things. The geologist studies the earth. The physicist is involved with energy in all of its forms. The chemist tries to understand the material substances that we find on earth and everywhere in the universe. The chemist often copies from nature to create new substances that are often superior to and cheaper than natural materials. We try to make nature serve us. Without science and technology, we are at nature's mercy. To control nature, we must learn its laws, then use them as we must.

This book is designed to extend your knowledge and understanding of your material universe. You will see how chemistry is changing and improving your life. You will be introduced to new and imaginative career fields that you may not have been aware of. These fields are in foods, plastics, cleansing agents, energy forms, environmental pollution, metallurgy, and poisons, to name a few. There are career opportunities for you in the science area. By the end of the term, you may want to know more about how you can participate in the improvement of our lives through science. This course is your first step in that direction. Speak with your chemistry teacher and guidance counselor to plan further steps that will prepare you for entry into many important and exciting career fields. Regardless of your career plans, this course will make you a far more aware and well-informed citizen in the most advanced scientific society in human history. Your votes and decisions will determine the directions of world peace, personal health, and prosperity.

1-1   How Do Scientists Solve Problems?

Decisions are the solutions to problems. They are being made at all times by everyone. Scientists, business executives, and high-ranking governmental officers all have a systematic approach to the solution of problems. In business management, problem solving is called "decision making." In the sciences, the process is called **the scientific method**. If you can teach yourself to apply and master the steps of the scientific method, you will hold a key that can unlock the door to success in any career field that you choose. This system involves the following five steps.

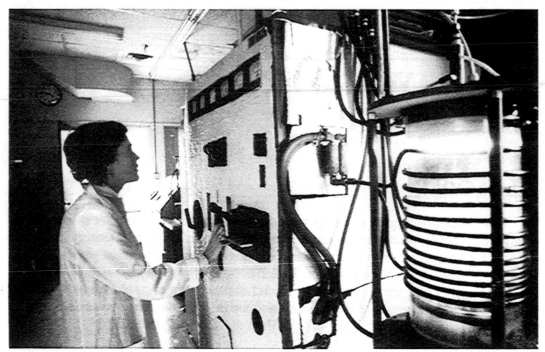

Courtesy: Socony-Mobil Oil Co.

FIGURE 1-1a   Top

FIGURE 1-1b   Left

Career opportunities in chemistry cover many areas of interest.

Courtesy: DuPont Co.

The Steps of the Scientific Method
(A System for Solving Problems and Making Decisions)

STEP 1: DEFINE THE PROBLEM   There can be no solution until a problem exists and is clearly recognized. This first step is usually the most difficult one.

STEP 2: GATHER INFORMATION   The scientist derives information by observing nature. Information may also be gained from the library, or through speaking with experts. The scientist must investigate and gather correct information before problems can be solved and conclusions made.

STEP 3: PROPOSE A HYPOTHESIS   **Hypothesis** is a compound of two Greek words. *Hypo* means "less than" and *thesis* means "idea." So a hypothesis is "less than an idea." It is a guess that might possibly serve to explain the gathered information. Using the hypothesis, the scientist can plan new ways to check his or her ideas.

STEP 4: EXPERIMENT   An **experiment** is a test to see if the hypothesis is correct. An experiment is an experience. Scientific experiments are **controlled**. A controlled experiment is regulated so only one factor is changed. The results of changing this factor are then compared with the unchanged event.

STEP 5: DEVELOP A THEORY   A **theory** is an idea. It is a tested conclusion. It is stronger than a hypothesis. Theories can be proven wrong and may be changed. People who are responsible for making decisions must be open-minded. Sometimes a theory becomes so well tested and accepted that we call it a **law**. Very few theories ever reach this position.

An Example of the Scientific Method in Action

STEP 1: PROBLEM: HOW DOES WINE SPOIL?   An answer to this problem was developed by chemist Louis Pasteur in the nineteenth century. He used the scientific method to achieve his results.

STEP 2: GATHER INFORMATION   The winemaking industry in France had a problem. Louis Pasteur was called upon to solve this problem. Some barrels produced spoiled, sour wine. He examined the sweet and sour wines under a microscope. He discovered germs in the sour wine that were absent in the sweet wine. He obtained this new information by careful observation.

STEP 3: HYPOTHESIS   Dr. Pasteur concluded that the newly discovered germs had caused the wine to turn sour.

STEP 4: EXPERIMENT   Louis Pasteur tested his hypothesis with a controlled experiment. He divided a barrel of sweet wine into two smaller casks. He introduced his newly discovered germs into one cask and left the other cask of sweet wine undisturbed. He observed that the wine with the added germs turned sour. The undisturbed wine remained sweet. He also observed that there were many more germs in the sour wine than he put into it.

STEP 5:   THEORY   Louis Pasteur concluded that his hypothesis was correct. If the experiment had proven the hypothesis wrong, Dr. Pasteur would have been required to return to step 2 and proceed again. His theory that germs caused the spoiling of wine had passed its tests. Theories must be tested *many* times. The more it is tested, the more validity the theory has. Louis Pasteur used his theory about wine spoilage to predict that diseases may be caused by germs. On investigation, he did discover germs associated with disease. **The germ theory of disease** has been verified by many scientists since Pasteur. The theory was revised because it was found that some diseases are not caused by germs.

1-2   How Can I Work Safely in the Lab?

You will be presented with scientific problems. You will use the chemistry laboratory to make observations (gather information) and formulate hypotheses. Your experiment may serve to verify a hypothesis. Chemistry is a laboratory science. You will be doing things and making things with chemicals. The lab can be a dangerous place because you will be working with glass, open flames, acids, poisonous chemicals, flammable liquids and gases. If you observe the safety rules of the lab, you will be safe and have a most rewarding experience.

1-3   What Are the Safety Rules for the Chemistry Lab?

1.   Do not enter the lab without your teacher present.

2.   Always work at your assigned place.

3.   Wear safety goggles and a lab apron when working in the lab.

4.   Tie back long hair.

5.   Lab tables should be clear of student belongings. The only things that should be on your lab table are your lab instructions, your pen, and the chemical materials provided by your teacher.

6.   Never perform an experiment without your teacher's permission.

7.   Do not use matches without your teacher's permission.

8.   Avoid playful, distracting, or boisterous behavior.

9.   Never eat or drink anything in the lab.

10.   Do not taste any chemicals.

11. If any chemical comes in contact with your skin, you should first wash it off under cool running water. Then you should report the accident to your teacher. Keep chemicals away from your eyes and face.

12. Rinse your hands after your experiment is completed, especially before lunch. You may have chemicals on your hands or under your fingernails.

13. Check all glassware for cracks. All glassware must be heat-resistant glass if it is to be heated.

14. When heating a test tube, use your test tube holder. Be sure that the open end is not pointing toward anyone.

15. Do not visit friends at another work station.

16. Do not pour reagents (chemicals) back into their bottles. Pour liquids down the drain or into the appropriate storage bottles as directed by your teacher.

17. Do not throw solid wastes into the sinks.

18. Clean up spills and accidents immediately.

19. Never leave a heating container unsupervised.

20. If you experience a problem with your experiment, stop and ask your teacher for help. Do *not* ask friends for help.

21. Report all accidents to your teacher.

1-4 How Should I Write My Lab Report?

All scientists write lab reports. In industry, the lab notebook becomes important in the establishment of commercial rights to a new product. In university research, the lab report is needed to support claims that may earn a Nobel Prize. The lab report should follow the steps of the scientific method. There are many different ways of writing up an experiment. Use the format your teacher will show to you. A suggested format for a lab report is as follows.

Lab Report Form

Name _____   Official Class _____

Chemistry Section _____   Date _____

Experiment #_____

Title _____

Problem _____

Procedure	Observation
1. _____	1. _____
_____	_____
_____	_____
2. _____	2. _____
_____	_____
3. _____	3. _____
_____	_____

Conclusions:

1. _____

2. _____

3. _____

Diagram

The following is an example of a student lab report.

Name ___John Jones___   Official Class ___11-3___

Chemistry Section ___15___   Date ___October 18, 19--___

Experiment #4

Title: The Metric System

Problem: To learn how metric units compare with English units.

Procedure

1. I drew a line ten inches long on the back of my lab sheet. I measured it with my metric ruler.

2. I filled a quart bottle with water and carefully poured the water into a 1-liter graduated cylinder.

3. I weighed a nickel on a scale.

Observation

1. The ten-inch line measured 25.4 centimeters (cm).

2. The water measured 946 milliliters (or 946 cc).

3. The nickel weighed about 5 grams.

Conclusions:

1. 2.54 cm equals 1 inch.
2. An inch is longer than a centimeter.
3. 1 quart is equal to 0.946 liters.
4. A liter is more than a quart.
5. 1 liter equals 1.06 quarts.
6. A nickel weighs about 5 grams.

Diagram:

1-5 How Did Different Systems of Measurement Come About?

Measuring systems are a requirement of civilization. The ancient Egyptians had to measure the land so each person's property could be easily identified. This was necessary because the Nile River overflowed every year and washed away the markings. Egyptian engineers needed accurate systems of measurement to build temples and pyramids. Their standard of length was the **cubit**, or the distance from your elbow to the end of your longest finger (see Fig. 1-2). The royal cubit was divided into two **spans** (or six palms or twenty-four finger widths). Check these measurements against your own cubit. This system of measurement was convenient for the Egyptians.

The measuring system used in the United States is called **the English system**. This system is complicated because of the irregular relationships the units have to each other. The basic unit of length is the **foot** (heel to toe). The foot is divided into twelve **inches**. A **yard** is the distance from the nose to the end of the arm. An inch is the width of an average thumb (see Fig. 1-3). The inch is also the length of four barley grains laid out in a line. The basic unit of weight is the **pound**, which is divided into sixteen **ounces**. The basic unit of volume is the **gallon**, which comprises four **quarts**. There does not seem to be any easy division in the English system of measurement.

The system most often used all over the world today is the **metric system**. This system is the easiest to work with. The metric system originated in France in 1798. Napoleon I, the Emperor of France at the time, welcomed and rewarded scientific achievement. The metric system is one of the many ideas that he encouraged.

1-6 Why Is the Metric System Used All Over the World?

The metric system is simple. It has very few units to learn. Larger or smaller units are easily identified by the prefix before its name. All units are in multiples of ten. Multiplication and division by ten is easy because it only requires the shifting of a decimal point. There is less chance for error when using the metric system.

FIGURE 1-2

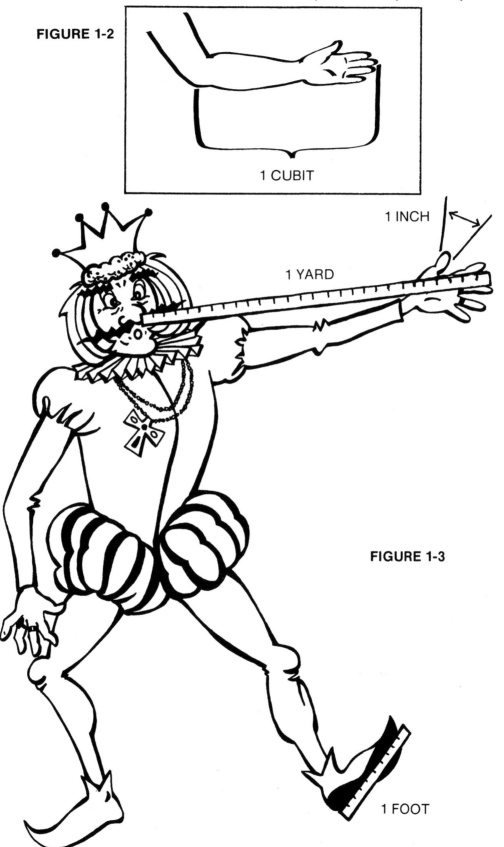

1 CUBIT

1 INCH

1 YARD

FIGURE 1-3

1 FOOT

Table 1-1 The Prefix System of Metric Units of Measurement

Prefix	Meaning	Symbol		Example	(length)	
mega-	million	M	Mm	megameter	=	1,000,000 meters
kilo-	thousand	k	km	kilometer	=	1,000 meters
hecto-	hundred	h	hm	hectometer	=	100 meters
deka-	ten	da	dam	dekameter	=	10 meters
			m			1 meter
deci-	one tenth	d	dm	decimeter	=	1/10 meter
centi-	one hundredth	c	cm	centimeter	=	1/100 meter
milli-	one thousandth	m	mm	millimeter	=	1/1,000 meter
micro-	one millionth			micrometer	=	1/1,000,000 meter

The *decimeter* is about the width of a person's hand. The *centimeter* is about the width of the nail on your little finger. The *millimeter* is about the thickness of your thumbnail.

Table 1-2 Common Abbreviations Used in the Metric System

Length		Mass (Weight)*		Volume	
kilometer	km	kilogram	kg	liter	l
meter	m	gram	g	milliliter	ml
centimeter	cm	centigram	cg	cubic centimeter	cc
millimeter	mm	milligram	mg	(cc = ml)	

*The gram and kilogram are units of mass. They are also often used as units of weight. This common practice will be used in this text.

The metric unit of volume is the **liter**. A liter is the space occupied by a cube that is 10 cm long on each edge. An added advantage of the metric system is that one liter of water at 4°C has a mass of one kilogram, so one ml of water has a mass of one gram. This fact gives us a connection between the units of volume and mass (Fig. 1-4).

FIGURE 1-4

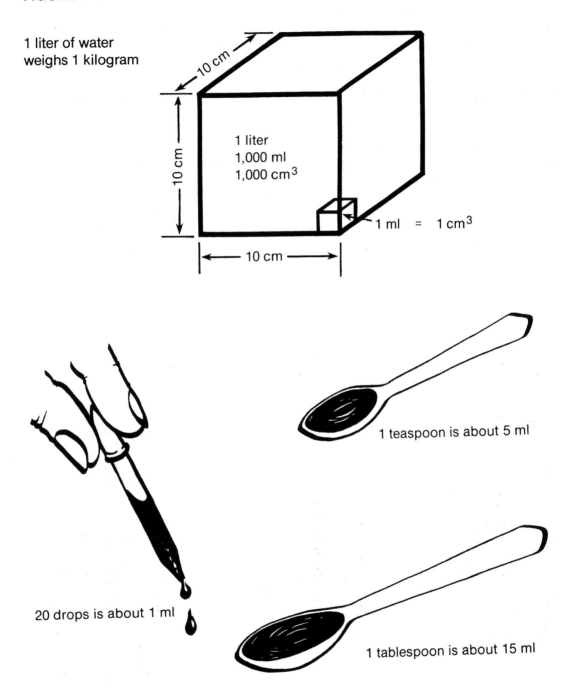

1 liter of water
weighs 1 kilogram

10 cm

10 cm

10 cm

1 liter
1,000 ml
1,000 cm^3

1 ml = 1 cm^3

20 drops is about 1 ml

1 teaspoon is about 5 ml

1 tablespoon is about 15 ml

1-7 How Is Temperature Measured?

A **thermometer** is an instrument used to measure temperature. Many thermometers contain the liquid metal mercury inside a hollow glass tube. This metal, like most materials, expands when it is warmed and contracts when it is cooled. The temperature scale can be made by marking the location of the mercury in the glass tube at two set temperatures. These two temperatures are the freezing point and the boiling point of water. The glass tube is then marked off in equal divisions or **graduations**. In the metric system, there are one hundred graduations between the freezing point and boiling point of water. It is for this reason that the metric temperature scale is called the **centigrade** scale. **Centi** means one hundredth, and **grade** means graduations or equal divisions. Another name for the centigrade temperature scale is the **Celsius scale**. It is named after **Anders Celsius**, the Swedish astronomer who invented it in 1742.

The English system of measurement uses the Fahrenheit temperature scale. It is not as simple as the centigrade scale. On the Fahrenheit scale, the boiling point of water is 212°F and its freezing point is 32°F. The scale is then divided into 180 graduations. Gabriel Daniel Fahrenheit invented the alcohol thermometer in 1709 in England. He also invented the mercury thermometer in 1714. The zero point on the Fahrenheit scale was the temperature of a mixture of equal parts of salt and snow. The salt lowered the freezing point of water (see section 5-8) from +32°F to 0°F (Fig. 1-5).

FIGURE 1-5

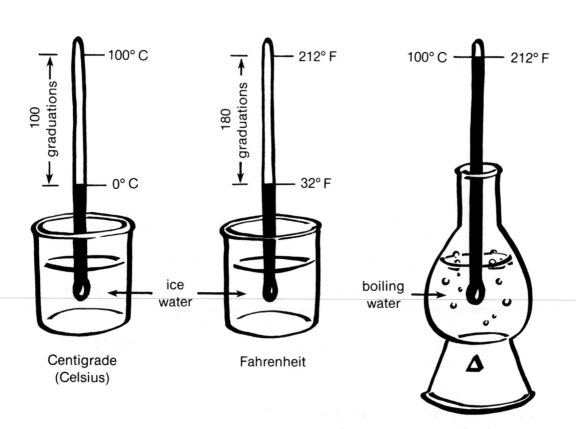

Centigrade (Celsius)   Fahrenheit

Temperature Scales

$$F = \frac{9}{5} C + 32$$

$$C = \frac{5}{9} (F - 32)$$

Table 1-3

Table of Metric and English Conversion Units

2.54	cm	=	1 inch		1	kg	=	2.2 pounds
1	m	=	39.37 inches		1	l	=	1.06 quarts
1	km	=	0.62 miles		3.79	l	=	1.0 gallon
28.35	g	=	1.0 ounce		28.3	l	=	1 cubic foot
453.6	g	=	1 pound		0.4536	kg	=	1 pound

Now You Know

1. Science is the systematic study of nature.

2. Chemistry is the study of matter or the material substances in the universe.

3. The field of chemistry offers many career opportunities.

4. The scientific method is a systematic way to solve problems or make decisions.

5. The chemistry lab is an important work area for the chemist. If the safety rules are observed, you will have a safe and rewarding experience.

6. All laboratory experiences should be written up in a lab report.

7. The metric system is used as the standard form of measurement in all parts of the world except the United States. The system is used by all scientists.

8. All metric units are in multiples of ten. This minimizes arithmetic operations and the chances of arithmetic errors.

9. The metric unit for length is the meter.
   The metric unit for weight (mass) is the gram.
   The metric unit for volume is the liter.
   Temperature is expressed in Celsius or centigrade degrees.

New Words

Celsius	The name of the metric temperature scale.
centigrade	Another name for the metric temperature scale.
chemistry	The study of the substances composing our universe.
controlled experiment	An experiment in which one factor is changed. The results are then compared with the event where no factors were changed (the control).
experiment	A test to confirm a hypothesis or a theory. An experiment is an experience.
gram	The unit of mass in the metric system. Also used as a unit of weight in common practice.
hypothesis	A statement that may be tested by scientific methods.
liter	The metric unit of volume.
metric system	The measuring system used by scientists.
science	The study of nature and natural events.
scientific method	A systematic approach to the solution of problems.
theory	An idea confirmed by experiment.

Metric Math Problems

Equating volume with the weight of water.
A special metric bonus!

(Note that 1 ml of water weighs 1 gram at 4° C
and 1 ml is the same as 1 cm³.)

Example Problem

A glass fish tank measures 30 cm long,
20 cm in height, and 10 cm wide.

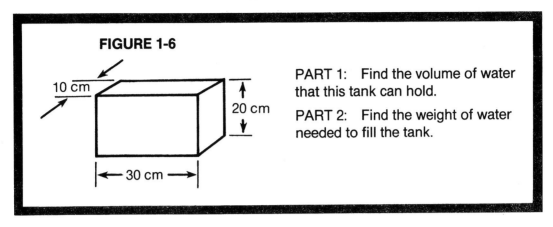

FIGURE 1-6

10 cm

20 cm

30 cm

PART 1:   Find the volume of water that this tank can hold.

PART 2:   Find the weight of water needed to fill the tank.

Solution:

PART 1   The volume of the tank is calculated by multiplying its length by its width, and by its height.

$$\text{volume} = \text{length} \times \text{width} \times \text{height}$$
$$\text{volume} = (30 \text{ cm}) \times (10 \text{ cm}) \times (20 \text{ cm})$$
$$\text{volume} = 6{,}000 \text{ cubic centimeters } (6{,}000 \text{ cm}^3)$$

Answer: This tank can hold 6,000 cm^3 of water at 4°C.

PART 2   Since 1 cm^3 of water weighs 1 gram, 6,000 cm^3 of water would weigh 6,000 grams (or 6 kilograms).

Metric System Problems

1. What is the weight, in grams, of 750 ml of water at 4°C?

2. Find the metric volume of 55 g of water at 4°C.

3. If a chemist added 35 g of saltwater to 165 ml of sugar water, how much would the resulting solution weigh? About how many milliliters would the resulting solution occupy?

4. If a box measures 10 cm in length, 5 cm in width, and 3 cm in height, about how many grams of water would be needed to fill the box?

5. If the empty box weighed 100 g, how much would the box weigh when filled with water?

6. An empty box weighs 400 g. When it is filled with water it weighs 1,600 g. Estimate the volume of this box.

7. If the box in problem 6 is 5 cm long and 2 cm wide, how high is it?

8. An empty thermos bottle weighs 350 g. When filled with tea, it weighs 1,050 g. About how many milliliters of liquid can this thermos hold?

9. If one teaspoon holds 5 ml of tea, how many grams of water can it hold?

10. How many teaspoonfuls of water are needed to fill a one-liter flask? (Ask your teacher if you can check your answer in the lab.)

Reading Power

For each of the following questions, select one answer that seems most correct.

1. The main idea of the introduction to this chapter is:

   a. Learn about our material universe and be introduced to new career fields.
   b. Science has many interesting career fields available to you.
   c. The United States is the most advanced technological society in the world.
   d. Expand our knowledge of the universe.

2. The main idea of section 1-1 is:

   a. The scientist has a system for solving problems that is different from that of business executives, military leaders, and high-ranking governmental leaders.
   b. The scientific method requires an experiment.
   c. The scientific method requires at least five steps.
   d. Often, the most difficult step in the scientific method is to define the problem.

3. The main idea of section 1-2 is:

   a. You will solve scientific problems in the lab.
   b. The main purpose of the chemistry lab is to gather information.
   c. The chemistry lab is a dangerous place to work.
   d. You will be safe in the chemistry lab if all safety rules are followed.

4. The main idea of section 1-4 is:

   a. Scientists in industry and in research write lab reports.
   b. Your lab report should follow the format your teacher describes.
   c. The lab report should follow the steps of the scientific method as directed by your teacher.
   d. There are two ways to write a lab report.

5. The main idea of section 1-5 is:

   a. Napoleon paid an award to the inventor of the metric system.
   b. The English system is equally as accurate as the metric system.
   c. Civilization requires an accurate system of measurement.
   d. The English system is difficult and complicated.

6. The main idea of section 1-6 is:

    a. The metric system is simple to use.
    b. The metric unit of length is the meter.
    c. The metric unit of mass is the gram.
    d. Prefixes before each metric unit tell the size of the unit.

7. The main idea of section 1-7 is:

    a. There are one hundred graduations between the boiling point and the freezing point of water on the centigrade scale.
    b. The mercury thermometer was invented by Gabriel Daniel Fahrenheit.
    c. Metric temperature is measured on the centigrade or Celsius temperature scale.
    d. The thermometer is used to measure temperature.

Mind Expanders

1. How is science different from the following subjects?

    a. English   b. social studies   c. mathematics

2. Why is a knowledge of science important for effective citizenship?

3. What is the scientific method?

4. Outline the steps of the scientific method.

5. How can you apply the scientific method to:

    a. improve your grade in chemistry?
    b. improve your grades in other subjects?

6. Why is the metric system preferred over the English system?

The Scientific Method

Arrange the following activities in their proper order for the solution of a problem according to the scientific method.

1. Observe pictures of the earth taken from space.

2. Is the earth round or flat?

3. I have confirmed my suspicion that the earth is round from all the evidence available.

4. Observe the earth's shadow as it moves across the moon during an eclipse of the moon.

5. This piece of evidence leads me to suspect that the earth may be round.

True or False

If the statement is true, mark it *T*. If the statement is false, change the word(s) in italics, using the list below, to make the statement true. You may use the same word(s) more than once.

metric	English
make observations and tests	equally as
theory	controlled
chemistry	scientist
Gabriel Daniel Fahrenheit	

1. The *mathematician* devotes his life to understanding the mysteries of nature.

2. A scientific *theory* can be proven to be true.

3. The science laboratory is a place to *define problems*.

4. The metric system is *more* accurate than the English system.

5. The alcohol and mercury thermometers were invented by *Gabriel Daniel Fahrenheit*.

6. *Chemistry* is the study of the materials that compose our universe.

7. A *hypothesis* is an idea that is confirmed by experiment.

8. The *English* system is used by American scientists.

9. The Fahrenheit temperature scale is part of the *English* system.

10. A *controlled* experiment is one in which one factor is changed.

Find the Facts

For each true/false question answered, locate the section in the text that supports your answer. Write the section number in your notebook next to your answer.

Multiple-Choice

1. Chemistry is the study of:

    a. living things   b. the earth   c. energy   d. matter

2. The scientific method includes all of the following except:

    a. imaginative thought   b. revision

    c. retesting   d. superstitious belief

3. All of the following are metric units of measure except:

    a. meter   b. cubit   c. liter   d. gram

4. The following unit of measure is the same for the English and metric systems.

    a. length   b. weight   c. time   d. gram

5. Which of the following is the longest unit of length?

    a. micrometer   b. dekameter

    c. hectometer   d. kilometer

6. Which of the following is the smallest unit of time?

    a. microsecond   b. dekasecond

    c. hectosecond   d. kilosecond

7. Science is the study of:

    a. geometry   b. human history

    c. natural events   d. foreign languages

8. The metric unit for mass is:

    a. pound   b. ounce   c. liter   d. gram

9. The metric temperature scale is:

    a. Celsius or Fahrenheit   b. Celsius or centigrade

    c. Fahrenheit or centigrade   d. Fahrenheit alone

10 The metric system is used as a standard by all countries except:

    a. France   b. Mexico   c. Italy   d. the U.S.A.

Lab Safety Test

Are you ready to work in the chemistry lab?

If the statement is true, mark it *T.* If the statement is false, eliminate or change the word(s) in italics to make it true.

1. You may enter the lab if your teacher is *not* present.

2. You *must* work at your assigned station in the lab.

3. It is important to wear *safety goggles* and a lab apron in the lab.

4. Long hair *may be* worn loose in the lab.

5. Do not store your books and coats on the *lab table* where you are working.

6. *Always* perform your experiment without your teacher's permission.

7. Do *not* use matches in the lab without your teacher's permission.

8. The lab is an *unsafe* place for playful and boisterous behavior.

9. *Always* eat and drink in the lab.

10. If chemicals get into your eyes, first flush your eyes with water, *then* tell your teacher.

11. There is *no* need to wash your hands after each lab, especially before lunch or dinner.

12. You may *use cracked* glassware to heat liquids.

13. When heating a liquid in a test tube, you should *never* point the open end of the test tube at anyone.

14. You may *never* visit friends at their work station to discuss the experiment without your teacher's permission.

15. *Never* pour excess chemicals back into the reagent bottle.

16. Throw all *solid* wastes into the sinks.

17. Do *not* clean up any spills. Leave them for the next class.

18. *Always* watch heating liquids or solids in the laboratory.

19. *Never* ask your teacher for help in the lab. Ask your friends instead.

20. *Never* report any accidents to your teacher.

Metric-English Conversion Problems

Use the table of metric-English conversion units to compute the values shown below.

2 inches	= _____ cm		20 kg	= _____ lbs
226.8 g	= _____ lb		10 lbs	= _____ kg
10 liters	= _____ quarts		10 ounces	= _____ g
10 km	= _____ miles		4 gallons	= _____ liters

Careers in Science

2-Year Degree	4-Year or Graduate Degree
lab technician	chemist
drug technician	pharmacist
dental lab technician	dentist
medical lab technician	medical doctor
practical nurse	registered nurse
paramedic	physician's associate

Chapter 2

How Does Matter Change?

Instructional Objectives

After completing this chapter, you will be able to:

1. Define amorphous, freezing point, gas, kinetic molecular theory, liquid, melting point, molecule, solid, temperature, and cite one example of each.

2. Define states of matter by their properties, and cite examples of each state of matter.

3. Distinguish (compare) physical and chemical changes and cite examples of each.

4. Explain phase changes via the kinetic molecular theory.

5. Recall the freezing point and boiling point of water.

Chapter 2 Contents

Introduction 27

2-1 What Is Matter? 27

2-2 How Does Matter Show Itself to Us? 29

2-3 What Kind of Changes Can Matter Undergo? 29

2-4 What Is Matter Made Of? 31

2-5 How Is Energy Involved with Molecules? 32
    (The Kinetic Molecular Theory)

2-6 How Does Matter Change Its State? 32

Written Exercises 37

Chapter 2

How Does Matter Change?

Introduction

Look around your room. There are two different kinds of things in this room. These two things are with you at all times. Your own body is made of these two different things. You and the room you are in are a very small part of the vast universe of planets and stars and they are all made of these two different things. Can you tell what they are? These two things are **matter** and **energy**. Energy comes in many forms. We know it as heat, light, electricity, and motion, to mention only a few of its many forms. In chemistry, we will learn about energy later, but our main purpose now is to understand matter.

2-1   What Is Matter?

How do you distinguish one friend from another? Of course, each person looks different from another, has a different voice, and behaves differently. Each person is recognized by his or her own characteristics. Matter also has characteristics or properties that allow us to recognize it. *All matter on the earth has weight, and all matter takes up space.* We often purchase material objects by their weight or by the space they occupy. Milk is measured by the quart and meat is measured by the pound. Matter is measured by making use of its two properties. Anything that has the properties of having weight and taking up space *must* be matter.

2-2 How Does Matter Show Itself to Us?

Matter takes three forms or states: **solid**, **liquid**, and **gas**. Each state is distinguished by its properties.

Table 2-1		
Properties of Solids	**Properties of Liquids**	**Properties of Gases**
Occupies a definite space	Occupies a definite space	Occupies any space
Has a definite shape	Has no shape	Has no shape

FIGURE 2-2

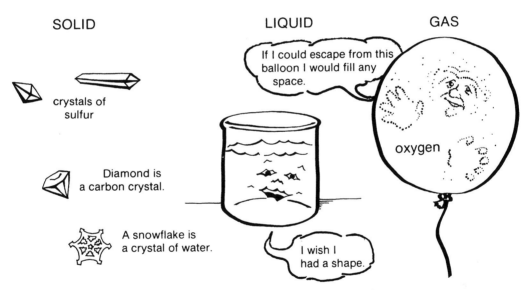

Liquids and gases have no shape of their own. They take on the shape of the container they are in. In weightless space, liquids that are not enclosed in a container take the shape of a ball (sphere). In addition, gases will expand until they reach the walls of their container. They will occupy any space.

2-3 What Kind of Changes Can Matter Undergo?

When an artist carves a statue from a piece of wood, the wood takes on a new appearance. The value of the carved wood is increased as a result of its new look. *This kind of change, in which the substance remains the same, but only the appearance is different is called a* **physical change**. When wood is consumed in a fire,

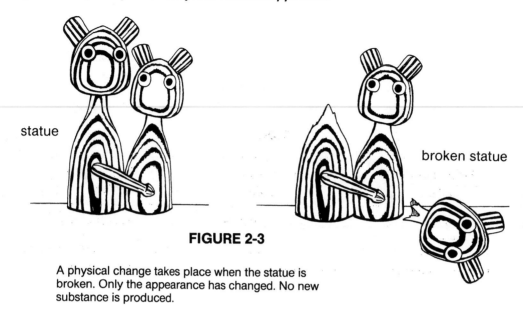

FIGURE 2-3

A physical change takes place when the statue is broken. Only the appearance has changed. No new substance is produced.

FIGURE 2-4

A chemical change takes place when the wooden statue is burned to ashes. A new substance is produced.

FIGURE 2-5

When sugar or salt dissolves in water it undergoes a change in appearance.

This is a PHYSICAL CHANGE.

FIGURE 2-6

What kind of change did this car undergo? Explain.

ashes result. This change of wood into ashes is called a **chemical change.** *In a chemical change a new substance is produced.* Wood has the property of being able to burn. Ashes cannot burn.

2-4   What Is Matter Made Of?

If you cut a cube of sugar in half, you would get two smaller pieces of sugar. If this process were continued, the pieces of sugar would get smaller. Eventually, you would arrive at the smallest particle of sugar that could exist, a **molecule** of sugar. This molecule is the smallest particle of sugar that can be identified as sugar. *All matter—solids, liquids, and gases—are made of molecules.* We can see that molecules are held together by attractive forces because substances do not fly apart. The molecules of a solid are packed more closely together and have little freedom of motion. In liquids, molecules move with more freedom and are able to flow. The molecules of gases have the greatest degree of freedom and their attractive forces are unable to hold them together.

2-5   How Is Energy Involved with Molecules? (The Kinetic Molecular Theory)

Kinetic is the Greek word that means "motion." **Theory** is the Greek word for "idea." *The kinetic molecular theory states that all molecules are always moving.* When you open a bottle of ammonia, you can smell the fumes all over the room. These gaseous molecules move from place to place. If you warm the ammonia solution, its fumes will spread faster. The kinetic molecular theory says:

a. Molecules never stop moving.

b. At higher temperatures, the molecules move faster.

c. Temperature is a measure of the speed with which molecules move.

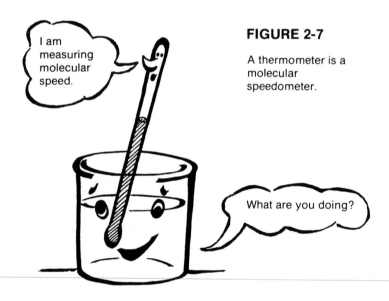

FIGURE 2-7

A thermometer is a molecular speedometer.

I am measuring molecular speed.

What are you doing?

2-6   How Does Matter Change Its State?

The molecules of a solid are most tightly held to each other and most closely packed together. If you could see these molecules, they would appear to be held together on a vibrating spring.

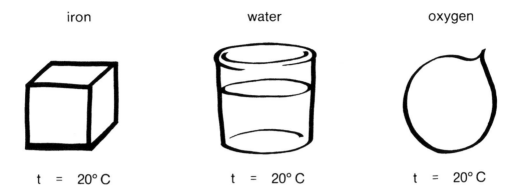

FIGURE 2-8

Why do the molecules of the iron, water, and oxygen all have the same kinetic energy?

If heat were added, the temperature would rise and the molecules would vibrate faster. At a certain temperature, the vibrations would strain their attractive forces to their limit. The addition of any more heat would result in the breaking of their connections. The result is that the solid would melt to a liquid. The temperature at which solids change into liquids is called the **melting point**.

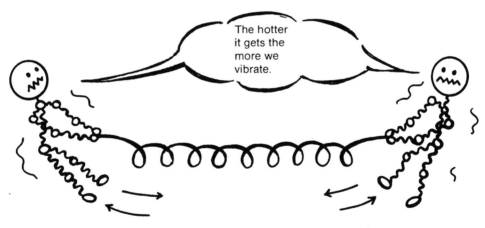

FIGURE 2-9

Molecules in the solid state seem to be held together on a vibrating spring.

When you put water into your freezer, the water temperature falls. The liquid molecules slow down. The attractive forces cause the molecules to lock together into the solid state. The change of a liquid into a solid is called **freezing**. The temperature at which freezing takes place is the **freezing point**. Since the molecules lock and unlock at the same temperature, the melting point and the freezing point are the same temperature. Why is freezing water into ice a physical change?

The molecules of a liquid have more freedom of motion than those of a solid. Molecules of liquids are also strongly held to each other by attractive forces. If a liquid is warmed, its molecules will move faster. As molecules gain energy, some will break free of their bonds. They escape from the liquid to become a gas, or vapor. Molecules in the gaseous state have the most freedom. **Evaporation** is the change of a liquid into a vapor.

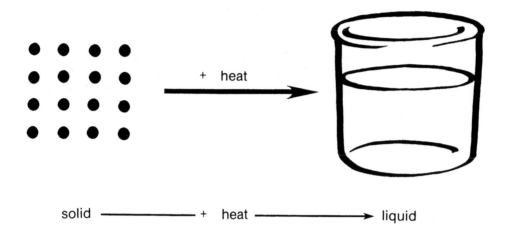

solid ———————— + heat ————————→ liquid

FIGURE 2-10

Since no new substance is produced, but the appearance has changed, melting is a physical change.

Condensation is the change of a vapor into a liquid (Fig. 2-10). It is the *opposite* of evaporation. As the temperature falls, and the gaseous molecules slow down, their weak attractive forces get an opportunity to bind the molecules together, and change the gas (vapor) into a liquid. When water vapor touches cool dust particles in the air, condensation takes place. The droplets of water, suspended in the air, form clouds. These droplets may fall to the ground as rain (see section 13-5a).

GAS ———————— Condensation ————————→LIQUID

When you put mothballs into your closet at home, they grow smaller each day and eventually disappear. The mothballs are solid, but have a strong odor. You can only smell a substance when it is in a gaseous state. The solid mothballs slowly turn to vapor. The changing of a solid into a gas (without becoming liquid) is called **sublimation**. A solid will melt when enough heat is added to break the attractive forces between the molecules which make up that solid. When sublimation takes place, much more heat is added to the solid. This added heat causes the molecular vibrations to become so violent that the molecules of the solid completely break away from each other and enter into a gaseous state.

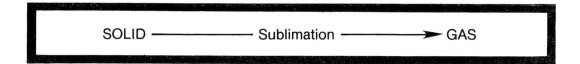

SOLID ——————— Sublimation ——————→ GAS

We know that water vapor will condense on a cool speck of dust. If the water vapor touches a very cold speck of dust in the air, the gaseous water may **crystallize** without condensing first. The result is a beautiful snowflake, or ice crystal. These ice crystals, suspended in the air, form clouds. If conditions are right, these crystals may fall to the ground as snow. The changing of a gas into a solid is also called *sublimation*, or *crystallization*.

GAS ——————— Sublimation ——————→ SOLID

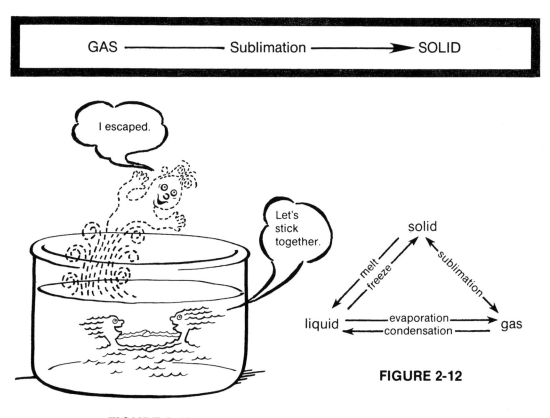

FIGURE 2-11

The changing of a liquid into a gas is called *EVAPORATION*.

FIGURE 2-12

Now You Know

1. Matter and energy are everywhere at all times.

2. Everything is identified by its properties (appearance and behavior).

3. Matter has two outstanding properties. All matter on the earth has weight and takes up space.

4. There are three states of matter: solid, liquid, and gas.

5. Matter undergoes two kinds of changes: **physical changes** and **chemical changes**.

6. A physical change involves a change in appearance only, and no new substance is produced.

7. A chemical change is a change in which a new substance results.

8. Matter is made of molecules.

9. Molecules are always moving.

10. Molecular motion is measured as temperature.

11. The motion of the molecules in the solid state are confined to vibrations.

12. Melting is the change of a solid into its liquid state.

13. Freezing is the change of a liquid into its solid state.

14. Condensation is the change of a gas into its liquid form.

15. Evaporation is the change of a liquid into its gaseous state.

16. Sublimation is the change of a gas into its solid state and vice versa.

New Words

amorphous	Shapeless.
freezing point	The temperature at which a liquid changes into a solid.
gas	Matter that does not occupy a definite volume, and is shapeless (amorphous).
kinetic molecular theory	The idea that molecules are in constant motion.

liquid	Matter that occupies a definite volume, but is shapeless (amorphous).
melting point	The temperature at which a solid changes into a liquid.
molecule	The smallest particle of a substance that can exist with its own identify.
solid	Matter that occupies a definite space (volume) and has a definite shape.
temperature	A measure of molecular motion.

Reading Power

For each of the following questions, select one answer that seems most correct.

1. The main idea of the introduction is:

    a. Everything contains both matter and energy.
    b. We are a part of the universe.
    c. Everything is made of energy.
    d. Energy has many forms, and we know it as heat, light, electricity, and motion.

2. The main idea of section 2-1 is:

    a. You can recognize your friends by their characteristics.
    b. Matter is measured by its properties.
    c. All things have their own properties.
    d. Matter occupies space and has weight.

3. The main idea of section 2-2 is:

    a. Solids occupy a definite space and have a definite shape.
    b. Liquids occupy a definite space and have no shape.
    c. Matter exists as a solid, liquid, and gas.
    d. Gases do not occupy a definite space, nor do they have any shape.

4. The main idea of section 2-3 is:

    a. A broken vase or statue is an example of a physical change.
    b. A physical change does result in a new substance.
    c. A chemical change does not result in a new substance.
    d. Matter can undergo both physical and chemical changes.

5. The main idea of section 2-4 is:

    a. Matter is made of tiny pieces of sugar.
    b. Sugar is made of molecules.
    c. All matter is either solid, liquid, or gas.
    d. All matter is made of molecules.

6. The main idea of section 2-5 is:

    a. Ammonia fumes will spread throughout the room.
    b. Molecules move faster at lower temperatures.
    c. Temperature is a measure of molecular motion.
    d. There is no relationship between energy and matter.

7. The main idea of section 2-6 is:

    a. Matter can change from one state to any other state by changing temperature.
    b. Matter can only exist in one state at a time.
    c. The freezing point and melting point are the same temperature.
    d. The molecules of solids, liquids, and gases at the same temperature have the same kinetic energy.

Mind Expanders

1. Explain why the evaporation of water is a physical change.

2. Write down two examples of physical changes that were not mentioned in this chapter.

3. How are solids and liquids similar? How are they different?

4. How are liquids and gases similar? How are they different?

5. How is the kinetic energy of a gas different from that of a solid?

6. How is the kinetic energy of a solid similar to that of a gas at the same temperature?

7. Explain how a thermometer measures molecular speed.

Physical and Chemical Changes

Identify each of the following as a physical or a chemical change:

1. Breaking your mother's favorite vase.

2. The freezing of water into ice cubes.

3. Burning a piece of paper.

4. Tearing of paper.

5. Dissolving instant tea in water.

Completion Questions

Choose the term from the list below that will correctly complete the following statements. (Some words may be used more than once.)

chemical	liquid
condensation	matter
does not occupy a definite space	melting
energy	molecules
evaporation	occupies a definite space
freezing	physical
gas	properties
has a definite shape	solid
has no shape	sublimation
kinetic molecular theory	temperature

1. The universe is made of _____ and _____ .

2. _____ occupies space and has weight.

3. All things are recognized by their _____ .

4. The three states of matter are _____ , _____ , and

   _____ .

5. The properties of the solid state are that it _____ and _____ .

6. The properties of the liquid state are that it _____ and _____ .

7. The properties of the gaseous state are that it _____ and _____ .

8. Matter can undergo a _____ change and a _____ change.

9. In a _____ change, no new substance is produced.

10. In a _____ change, there is only a change in appearance.

11. _____ is the changing of a liquid into a gas.

12. The idea that molecules are in constant motion is called the _____ _____ .

13. _____ is a measure of the motion of molecules.

14. _____ are the smallest particles of a substance that can exist.

15. _____ is the change of a solid into a liquid.

16. _____ are in constant motion.

17. When a liquid changes into a solid, it is called _____ .

18. _____ is the change of a gas into a liquid.

19. _____ is the opposite of evaporation.

20. The change of a gas into its solid is called _____ .

Matching

Write the number in column B of the statement that relates to the word(s) in column A in the spaces indicated.

COLUMN A	COLUMN B
_____ kinetic molecular theory	1. The disappearance of mothballs in a closet.
_____ sublimation	2. Measurement of molecular motion.
_____ freezing	3. Molecules are in constant motion.
_____ melting	4. A solid changing into a liquid.
_____ temperature	5. A liquid changing into a solid.

Chapter 3

How Is Matter Constructed?

Instructional Objectives

After completing this chapter, you will be able to:

1. Define atom, atomic nucleus, atomic number, atomic weight, electron, element, energy level, isotope, neutron, and proton.

2. Explain the Dalton atomic theory.

3. Recognize symbols and names of common elements.

4. Draw and label Bohr atomic diagrams (1–20).

5. Explain negative and positive charge in terms of electrons.

Chapter 3 Contents

	Introduction	43
3-1	What Are Substances Made Of?	43
3-2	What Is the Dalton Atomic Theory?	44
3-3	What Are Atoms Made Of?	44
3-4	How Are Atoms Structured?	45
3-5	How Is the Atomic Number Important?	46
3-6	How Is the Nucleus Important?	46
3-7	How Do the Electrons Orbit the Nucleus?	47
3-8	How Does the Periodic Table Reveal a Picture of the Atom?	48
3-9	How Can the Properties of the Elements Be Predicted by Trends in the Periodic Table?	49
3-10	How Do the Properties of the Elements Change in the Same Period?	51
	Written Exercises	54

Chapter 3

How Is Matter Constructed?

Introduction

In Chapter 2, we learned that the smallest part of a substance that can exist by itself is a molecule. If a molecule were to be taken apart, we would have the smaller particles of which it is composed. About 2,500 years ago, the Greek philosopher **Democritus** called these molecular parts "atomos." Today we call them **atoms**.

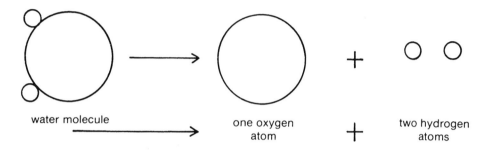

water molecule → one oxygen atom + two hydrogen atoms

The water molecule is made of three (3) atoms.

FIGURE 3–1

All molecules are made of atoms.

3-1   What Are Substances Made Of?

The ancient Greek philosophers believed that all the matter in the universe was composed of four elements: earth, air, fire, and water. Today we know of 109 different elements, and scientists are attempting to create additional elements in their laboratories. An **element** is a substance from which no other material can be obtained. Water is not an element because oxygen and hydrogen can be obtained from it. But oxygen and hydrogen are elements. Every element has its own **symbol**. These symbols are used throughout the world. They have the same meaning in China, the Soviet Union, the United States, and in all countries. These symbols form the words of a true international language because they have the same meaning for people all over the world. Study the symbols as they are listed in the table of the elements at the end of this book.

FIGURE 3-2

The ancient symbols of the elements.

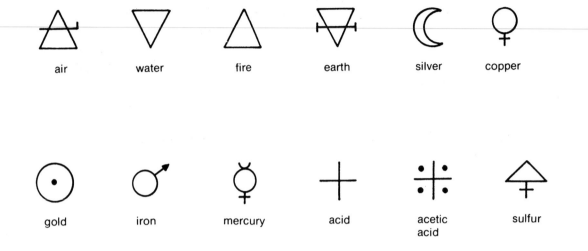

3-2   What Is the Dalton Atomic Theory?

In 1807 (about 25 years after the American Revolutionary War), English scientist **John Dalton** made the first meaningful progress in understanding the nature of matter since Democritus. The main ideas of Dalton's atomic theory are:

1. Atoms of the same element are the same.

2. Atoms of different elements are different.

3. Atoms combine to form new substances that are made of molecules.

4. Molecules of the same substance are made of the same kinds and numbers of atoms.

Dalton, like Democritus, was unable to tell what atoms were made of, how atoms were structured, or how they combined. All of these problems have been solved within the past 100 years.

3-3   What Are Atoms Made Of?

Rub your plastic pen on a piece of wool (or your sleeve), then bring it near a small snip of paper. The paper will be attracted to the rubbed pen. This is because the plastic pen acquired an electrical charge when it was rubbed against the wool. **Electrons** were rubbed from the cloth onto the pen. The electrons came from the atoms that make up the wool. The electron was discovered by the great scientist **J. J. Thompson** in 1897. Thompson showed, for the first time, that atoms are composed of particles and that one of these particles is the electron. He showed that electrons have very little weight and possess a negative electrical charge.

In 1911, **Ernest Rutherford** discovered that most of the mass of the atom was concentrated in a small positively-charged nucleus. He hypothesized that there were particles that carried a positive charge. He called these particles **protons**. The proton is about 2,000 times heavier than the electron and has a positive charge. In 1932, **James Chadwick** discovered the **neutron**. The neutron has about the same weight as the proton, but it carries no charge. The electron, proton, and neutron are the three most important particles composing atoms. Objects that have more electrons than protons have a negative charge. Objects that have fewer electrons than protons have a positive charge.

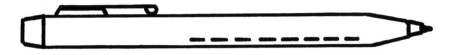

Excess electrons are rubbed off the wool onto the plastic pen.

snip of paper

FIGURE 3-3

The paper is attracted to the charged pen. Charged objects attract neutral (uncharged) objects.

3-4 How Are Atoms Structured?

Atoms are too small to be seen, but due to the hard work of many scientists, a picture of the atom has been revealed to us. In 1911, **Niels Bohr** suggested that electrons move about the atomic nucleus in definite orbits, or paths, similar to the way planets move around the sun. In the solar system, however, there is only one planet in each orbit, while in an atom there may be several electrons sharing the same orbit. Today we know that electrons are located in their own definite energy levels.

FIGURE 3-4

Atoms of the element sodium
have eleven (11) protons in
their nuclei, and twelve (12)
neutrons. Eleven electrons
orbit this nucleus.

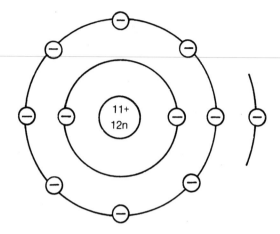

3-5 How Is the Atomic Number Important?

The number of protons in the nucleus is called the **atomic number**. The atomic
number defines the element.

Table 3-1

Atomic Number   =   number of protons

The atomic number determines what the element will be.

If the atomic nucleus contains one proton, then the element is hydrogen. Two
protons in the nucleus changes the element to helium, and eleven protons in the
nucleus can only mean the element sodium. Each element has its own atomic
number. Atoms of different elements cannot have the same number of protons in
their nuclei. Elements can be named by their atomic numbers. Atomic numbers
(proton numbers) could be part of a language we could use to communicate with
intelligent life from other planets or star systems. If we should discover such life in
distant space, communication using atomic numbers could enable us to identify
elements for each other.

3-6 How Is the Nucleus Important?

The nucleus of the atom is made of protons and neutrons. There are no electrons in the nucleus. The weight of the atom is concentrated in its tiny nucleus. The
atomic weight is the sum of all the protons and neutrons in the nucleus. Atoms

of the same element that have different numbers of neutrons also have different atomic weights. They are called **isotopes**.

Table 3-2

Mass Number = number of protons + number of neutrons

Table 3-3

	Charge	Mass Number	Location
Electron	−1	0	Orbits the nucleus
Proton	+1	1	In the nucleus
Neutron	0	1	In the nucleus

Atoms of the same element (same proton number) that have differing numbers of neutrons (different atomic weights) are called **isotopes**.

3-7   How Do the Electrons Orbit the Nucleus?

If you owned a tall apartment building that had no elevator, your tenants would want to occupy the apartments on the ground floor first. These apartments require the least amount of energy to be reached. As the lower floors fill up, the upper floors would become occupied. In an atom, the electrons also seek out the orbits that are closest to the nucleus, because they are located at a lower energy level. The low energy orbits fill first. The higher energy levels fill with electrons only after the lower energy levels are occupied. Niels Bohr labeled the lowest energy orbit as the **K shell**. The K shell is closest to the nucleus. The other orbits, or shells, are listed in alphabetical order: **K, L, M, N, O, P**, and **Q**. In the atomic diagrams shown, it can be seen that two electrons are needed to fill the K shell, eight electrons to fill the L shell, and, for light elements (atomic numbers 1–20), eight electrons will fill the M shell. It should be noticed that the number of orbiting electrons equals the number of protons (atomic number) in the nucleus.

Table 3-4

Element	Symbol	Atomic Number	Atomic Weight	Electrons			
				K	L	M	N
hydrogen	H	1	1	1–			
helium	He	2	4	2–			
lithium	Li	3	7	2–	1–		
beryllium	Be	4	9	2–	2–		
boron	B	5	11	2–	3–		
carbon	C	6	12	2–	4–		
nitrogen	N	7	14	2–	5–		
oxygen	O	8	16	2–	6–		
fluorine	F	9	19	2–	7–		
neon	Ne	10	20	2–	8–		
sodium	Na	11	23	2–	8–	1–	
magnesium	Mg	12	24	2–	8–	2–	
aluminum	Al	13	27	2–	8–	3–	
silicon	Si	14	28	2–	8–	4–	
phosphorous	P	15	31	2–	8–	5–	
sulfur	S	16	32	2–	8–	6–	
chlorine	Cl	17	35	2–	8–	7–	
argon	Ar	18	40	2–	8–	8–	
potassium	K	19	39	2–	8–	8–	1–
calcium	Ca	20	40	2–	8–	8–	2–

3-8 How Does the Periodic Table Reveal a Picture of the Atom?

Scientists always looked for order in the universe. As the elements were discovered, chemists searched for an orderly arrangement of the elements. In the nineteenth century, the Russian chemist Dmitri Ivanovich Mendeleev discovered that when the elements were listed by their atomic weights, the properties of the elements repeated themselves at regular intervals. He placed the elements with similar properties in a list, one under the other. Today we use this arrangement, but the elements are listed by their atomic numbers instead of their weights. This listing is known as the **periodic table**. It is periodic because the physical and chemical properties repeat themselves with regularity. The vertical columns list or **group** the families of elements. Each member of the **family** has similarities to other members of the same family. The horizontal lines are called **periods**.

At the time that the periodic table was created, scientists had not yet discovered the electron. They could not draw a diagram of the atom. They didn't realize that the periodic table was the "hidden" picture of the atom that they were searching for.

Look at the periodic table in this book (p. 381). Each period represents an electron shell. There are seven periods, for each of the seven electron shells found in all atoms. The elements in Period 1 have only one electron shell. This is their K shell, or the first electron shell, in the Bohr atom. The K shell can hold no more than two electrons. It is not by chance that there are two elements, hydrogen and helium, in Period 1. The elements in Period 2 have two electron shells. They are the K and the L shells (first and second electron shells). The L shell can hold eight electrons, and there are eight elements in Period 2. The elements in the third period have an M shell (the third electron shell). It can hold up to eighteen electrons. Eighteen elements are associated with Period 3, eight "regular" elements and ten transition elements (metals). The transition elements are found in Groups 3 through 12 on the periodic table. The elements in Period 4 have four electron shells. The fourth or N shell can hold thirty-two electrons when it is full. Thirty-two elements are associated with the fourth period (eight regular elements, ten transition elements, and fourteen lanthanides). Periods 5, 6, and 7 are also limited to thirty-two elements; however, Period 7 is incomplete, with only twenty-three elements.

The elements in Group 1 are similar in that they all have one valence electron. They are called the **alkali metals**. They are all extremely reactive metals. They must be stored under oil to protect them from burning up in the air. They all react violently on contact with water. In Group 2 all of the elements have two electrons in their valence shell. These elements are called **alkaline earth metals** because their metallic oxides feel earthy. These metals are also very reactive, but they are less active than the alkali metals of Group 1. Group 13 elements all have three valence electrons. Group 14 elements have four valence electrons, and so on. The transition metals are special. After the valence shell is occupied by two electrons, additional electrons fill the second-to-the-last electron shell, until it fills with eighteen electrons. The lanthanides and actinides fill the third outermost shell after the valence shell has two electrons. The transition metals, the lanthanide metals, and the actinide metals all make colorful compounds when they are dissolved in water.

3-9   How Can the Properties of the Elements Be Predicted by Trends in the Periodic Table?

As we go from one period to the next, another electron shell is added to the atom. This causes the atomic size to increase. The larger size means that the distance between the valence shell and the nucleus becomes greater. Additional electron shells cause more electrons to be located between the nucleus and the valence shell. We call these **shielding electrons**. The greater distance and the shielding electrons work together to weaken the attraction of the nucleus for its valence electrons. Atoms of **metals** have a weak hold on their valence electrons. Metals lose their valence electrons whenever they react. Active metals lose their valence electrons more easily than less active metals. Elements in the periodic table grow more metallic when they appear lower in the family or group because of their larger atomic size and their greater number of shielding electrons. We can predict that elements at the bottom of a group tend to become metallic.

Nonmetals are elements whose atomic nuclei have a strong attraction for their electrons. Their attraction is so strong that their atoms gain additional electrons when they react. The atoms at the top of the family are smaller and have fewer shielding electrons, so we find that the elements at the top of a group tend to be nonmetallic.

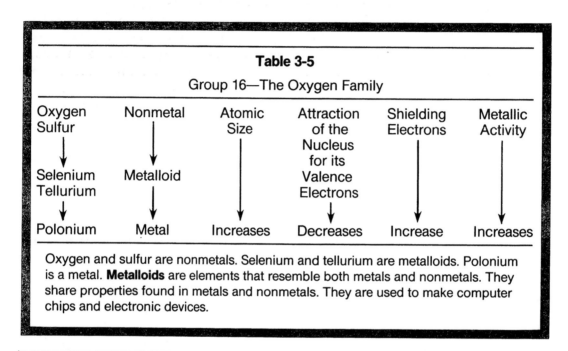

Table 3-5

Group 16—The Oxygen Family

Oxygen Sulfur ↓ Selenium Tellurium ↓ Polonium	Nonmetal ↓ Metalloid ↓ Metal	Atomic Size ↓ Increases	Attraction of the Nucleus for its Valence Electrons ↓ Decreases	Shielding Electrons ↓ Increase	Metallic Activity ↓ Increases

Oxygen and sulfur are nonmetals. Selenium and tellurium are metalloids. Polonium is a metal. **Metalloids** are elements that resemble both metals and nonmetals. They share properties found in metals and nonmetals. They are used to make computer chips and electronic devices.

Table 3-6

Group 17—The Halogen Family

		Color	State	Density
Fluorine	Nonmetal	Yellow	Gas	
Chlorine	Nonmetal	Green	Gas	↓
Bromine	Nonmental	Red-Brown	Liquid	
Iodine	Nonmetal	Grey	Solid	
Astatine	Metalloid	Grey-Black	Solid	Increases

Halogen is the Greek word for "salt generator" or "salt maker." A salt is formed when a halogen combines with a metal. *Chlorine* is the Greek word for "green." It derives its name from the fact that chlorine is a green gas under room conditions. *Bromine* is the Greek word for "stench," because of bromine's strong odor. *Iodine* means "violet" in Greek. Iodine vapor is violet, but iodine is grey in solid form. *Astatine* means "unstable." Astatine is a radioactive element. Radioactive elements are unstable because they often change into other elements.

The halogens, and all other families in the periodic table, increase their density as their atoms grow heavier. Dense objects have their molecules more closely packed together. The result is seen as the family changes from a gas to a liquid and finally to a solid. The color also deepens with the density. If the family were to continue beyond astatine, we would expect the next element to be a metal.

3-10 How Do the Properties of the Elements Change in the Same Period?

When we look along any period, from left to right, we see that the atomic number (see section 3-5) increases. The number of electron shells and shielding electrons remains the same throughout the period. As the atomic number increases, the more positive nucleus exerts a stronger attraction for its electrons. The electron shells are pulled closer to the nucleus. As a result, the atoms get smaller in size. The nucleus also has a stronger attraction for its valence electrons and doesn't release them easily. On the left side of the period, the elements are metallic. As we move to the right, the elements become metalloids, then nonmetals, and finally inert (nonreactive) gases.

Table 3-7

Period 2

Group	1	2	13	14	15	16	17	18
Element	Li	Be	B	C	N	O	F	Ne
Atomic Number	3	4	5	6	7	8	9	10

–Metals ——————→ Metalloids ———→ Nonmetals ———→ Inert Element

Atomic Number ————————————————→ Increases

Number of Electron Shells ————————————→ Remains the Same

Number of Shielding Electrons ——————————→ Remains the Same

Attraction of the Nucleus for the Valence Electrons ————————→ Increases

Atomic Size ————————————————————→ Decreases

Lithium and beryllium are metals. Boron and carbon are metalloids. Nitrogen, oxygen, and fluorine are nonmetals. Neon is an inert element.

There are six inert elements in the universe (helium, neon, argon, krypton, xenon [pronounced zee-non], and radon). There are only six nonmetals (fluorine, chlorine, bromine, iodine, oxygen, and sulfur). There are eleven metalloids in the entire universe (boron, carbon, silicon, germanium, nitrogen, phosphorous, arsenic, antimony, selenium, tellurium, and astatine). All of the remaining elements are metals.

Ninety-two natural elements are found on the earth. Those elements whose atomic numbers are greater than 92 are called **transuranium elements**. These elements are created by nuclear scientists. At this time we have seventeen transuranium elements. These new elements are all located in Period 7 in the periodic table. Their location tells us that they are all transition metals. About 80% of all the elements are metals.

Nuclear scientists are able to synthesize new elements, but each new element made can only be a metal.

Now You Know

1. All molecules are made of atoms.

2. All substances are made of elements or combinations of elements.

3. An element is a material from which no other substance can be obtained.

4. Dalton's atomic theory says:
    a. Atoms of the same element are the same.
    b. Atoms of different elements are different.
    c. Atoms of elements combine to form new substances.
    d. Each molecule of the same substance is made of the same kinds and numbers of atoms.

5. Atoms are made of electrons, protons, and neutrons.

6. Negatively-charged objects have an excess of electrons.

7. Positively-charged objects have a deficiency of electrons.

8. The number of protons in the nucleus is called the atomic number, or proton number.

9. The atomic number determines the element.

10. The mass number is the sum of the protons and neutrons in the nucleus.

11. Atoms of the same element that have different atomic weights are called isotopes.

12. The electrons move around the nucleus in definite energy levels as if they were in orbits.

13. The orbiting electrons fill the lowest energy levels first, then proceed to fill higher energy levels.

14. Two electrons fill the K shell, and eight electrons fill the L shell of all atoms. For light elements (atomic numbers 1–20), eight electrons fill the M shell.

15. The periodic table reveals a hidden picture of the atom.

16. The properties of the elements can be predicted by their position in the periodic table.

New Words

alkali metals	The elements in Group 1. They are very active metals that react with water to form strong bases.
alkaline earth metals	The elements in Group 2 in the periodic table. Their oxides crumble and feel like dry earth.
atom	The units of an element that make up molecules.
atomic nucleus	The central part of an atom that contains the protons and neutrons.
atomic number	The number of protons in the nucleus.
atomic weight	The net weight of all protons and neutrons in the nucleus.
electron	A particle that is a part of all atoms. It has a negative electrical charge and almost no weight.
element	Substances whose atoms have the same atomic number.
energy level	The orbit, or shell, in which the electron is located.
family	See *group* below.
group	A family (vertical list) of elements in the periodic table.
isotope	Atoms with the same atomic number, but a different atomic weight.
mass number	The sum of all protons and neutrons in the nucleus.
metal	An element whose atoms release valence electrons when they react.
metalloids	Elements that resemble both metals and nonmetals. They have properties found in metals and in nonmetals.
neutron	A particle that is a part of the atom. It has no electrical charge and has about the same weight as the proton.
nonmetal	An element whose atoms gain valence electrons when they react.
period	The horizontal listing of elements in the periodic table.
periodic table	The arrangement of all the elements by atomic numbers. Elements listed in vertical columns are in the same family because they share similar properties.

proton   A particle that is a part of all atoms. It has a positive electrical charge and weighs almost 2,000 times more than an electron.

shielding   Electrons located between the atomic nucleus and the val-
  electrons   ence shell.

transuranium   Elements whose atomic numbers are greater than 92. These
  elements   elements are made by scientists.

Reading Power

For each of the following questions, select one answer that seems most correct.

1. The main idea of the introduction is:
    a. The molecule is the smallest part of a substance that can exist by itself.
    b. Water is made of molecules.
    c. The atomic theory started with Democritus.
    d. Democritus lived about 2,500 years ago.

2. The main idea of section 3-1 is:
    a. All matter is a combination of earth, air, fire, and water.
    b. All matter is made of elements.
    c. Water is not an element because it can be broken down into new material.
    d. Every element has a symbol that is recognized in every country in the world.

3. The main idea of section 3-2 is:
    a. Dalton had no better understanding of how atoms are structured than Democritus.
    b. Dalton knew that atoms of the same element were the same but he did not understand how they were the same.
    c. Dalton knew that atoms combined to form molecules, but he did not understand how they did it.
    d. All of the above.

4. The main idea of section 3-3 is:
    a. All matter is neutral in nature.
    b. Molecules are positive.
    c. Atoms are made of electrical particles.
    d. Many scientists were involved in discovering what atoms are made of.

5. The main idea of section 3-4 is:
    a. Atoms are too small to be seen.
    b. One useful way of describing the atom is to compare its structure with that of the solar system.
    c. The solar system has only one planet in each orbit.
    d. The atom has more than one electron filling any orbit.

6. The main idea of section 3-5 is:
   a. The atomic nucleus is important.
   b. The atomic number defines the element.
   c. The atomic weight is the number of protons and neutrons.
   d. There are no electrons in the nucleus.

7. The main idea of section 3-6 is:
   a. The atomic weight is the number of protons and neutrons in the nucleus.
   b. The atomic number determines what the element will be.
   c. A nucleus of one proton must be hydrogen.
   d. Isotopes have the same proton number, but different neutron number.

8. The main idea of section 3-7 is:
   a. A tall apartment building with no elevator would have its ground floor apartments occupied first.
   b. The lowest energy level, or electron shell, is the K shell.
   c. Electrons fill the lower energy level orbits first.
   d. The proton number is the same as the atomic number.

9. The main idea of section 3-8 is:
   a. The periodic table is a systematic listing of all of the properties of the elements.
   b. The periodic table is a hidden diagram of the atom.
   c. The periodic table reveals the atomic number and the atomic weight of all of the elements.
   d. Eighty percent of the elements are metals.

10. The main idea of section 3-9 is:
    a. The location of an element in the periodic table allows us to predict the properties of the element.
    b. Metalloids fall between the metals and the nonmetals.
    c. No new natural elements can be discovered because they are all accounted for.
    d. Inert elements have complete valence shells.

11. The main idea of section 3-10 is:
    a. Atoms grow smaller as we move from left to right in any period.
    b. The properties of the elements in any period can be predicted by their location in the period.
    c. All atoms in the same period have the same number of shielding electrons.
    d. Metallic atoms can be found at the right-hand side of any period.

Mind Expanders

1. Explain, in your own words, how the symbols of the elements could be used as an international language.

2. What would the interstellar word for oxygen be?

3. What is the difference between an atom and a molecule?

4. Why is water *not* considered an element?

5. What evidence do we have that matter is electrical in nature?

6. Explain how atoms of lead are different from atoms of gold.

7. Why do many scientists say that atoms are mostly empty space?

8. How could two atoms of the same element have different atomic weights?

9. Why will all elements developed in the future be metals?

Complete the Following Table

Table 3-8

Element	Symbol	Element	Symbol
hydrogen		sodium	
	Li	aluminum	
boron			P
nitrogen			Cl
	F		K
iron		gold	
	Cu		Ag

Completion Questions

Using the words below, complete the following statements. (Some words may be used more than once.)

element	molecules	John J. Thompson
proton(s)	electron(s)	neutron(s)
different proton number(s)		

1. A material from which no new material can be obtained is a(n)_____ .

2. The atoms of different elements have _____ .

3. When atoms combine they form _____ .

4. The scientist who discovered the electron is _____ .

5. The _____ has no mass.

6. The number of _____ determines what the element will be.

7. The _____ has no electrical charge.

8. The _____ and _____ have equal masses.

9. The _____ has a negative charge.

10. The _____ has a positive charge.

Drawing Atomic Diagrams

In Table 3-9 on page 58, draw the atomic diagrams of each element in the box provided. The number below each symbol is the atomic number, and the number above the symbol is the atomic weight. Your drawings should conform to the sample given for K (potassium).

Table 3-9

$_1H^1$							
	Atomic Weight ⟶ 4 Atomic Number ⟶ 2 He						
$_3Li^7$	$_4Be^9$	$_5B^{11}$	$_6C^{12}$	$_7N^{14}$	$_8O^{16}$	$_9F^{19}$	$_{10}Ne^{20}$
$_{11}Na^{23}$	$_{12}Mg^{24}$	$_{13}Al^{27}$	$_{14}Si^{28}$	$_{15}P^{31}$	$_{16}S^{32}$	$_{17}Cl^{35}$	$_{18}Ar^{40}$
$_{19}K^{39}$ 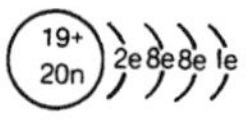	$_{20}Ca^{40}$						

19+ 20n) 2e 8e 8e 1e

Multiple-Choice

1. The sulfur atom, with 16 protons and 16 neutrons, has an atomic mass of:
   a. 16   b. 8   c. 32   d. 1

2. An atom with 6 protons and 8 neutrons in its nucleus is:
   a. carbon   b. oxygen   c. silicon   d. hydrogen

3. An atom with 12 protons and 13 neutrons must have _____ electrons.
   a. 6   b. 12   c. 13   d. 25

4. As the number of neutrons in the nucleus increases, the:
   a. atomic number increases
   b. element changes
   c. number of orbital electrons increases
   d. the atomic weight increases

5. The maximum number of electrons that can fill the L shell is:
   a. 2   b. 8   c. 18   d. 32

Answer questions 6 through 10 based on the atom that has an atomic number of 18 and an atomic weight of 40. Questions 11 through 15 are not about this specific atom.

6. The nucleus has:
   a. 40 protons and 18 neutrons
   b. 18 protons and 40 neutrons
   c. 22 protons and 18 neutrons
   d. 18 protons and 22 neutrons

7. The number of electrons in the K shell is:
   a. 2   b. 8   c. 18   d. 40

8. The number of electrons in the L shell is:
   a. 2   b. 8   c. 18   d. 40

9. The total number of electrons in all shells is:
   a. 2   b. 8   c. 18   d. 40

10. The neutron number of this atom is:
    a. 18   b. 22   c. 40   d. 8

11. When an atom loses two electrons, it acquires a charge of:
    a. $^-2$ b. $^+2$ c. $^-6$ d. $^+6$

12. The atomic particle with no electrical charge is:
    a. electron   b. proton   c. nucleus   d. neutron

13. The number of elements known to scientists is:
    a. 92   b. 96   c. 100   d. over 100

14. The atomic number is the number of:
    a. electrons
    b. protons and neutrons
    c. protons
    d. electrons, protons, and neutrons

15. Atoms of the same element have the same number of:
    a. electrons
    b. protons and neutrons
    c. neutrons
    d. protons only

Chapter 4

How Do Atoms Combine?

Instructional Objectives

After completing this chapter, you will be able to:

1. Define compound, covalent bond, element, formula, ion, ionic bond, metal, metallic bond, metallic luster, molecule, nonmetal, nonpolar, nonpolar covalent bond, opaque, polar, polyatomic ion, radical, subscript, valence, and valence electrons.

2. Compare metals with nonmetals.

3. Explain metallic properties in terms of metallic bonding.

4. Write the formulas of compounds given the symbols and valences of their elements and radicals.

5. Identify the elements (and radicals) in a compound from its formula.

6. Determine the number of atoms (and radicals) in a compound from its formula.

7. Calculate formula weight from the formula and the periodic table.

8. Calculate the percent by weight of each element in a compound from its formula and the periodic table.

9. Determine an empirical formula given a periodic table, and the percent by weight of the elements.

10. Define mole and Avogadro number.

11. Determine the number of moles of a substance from its weight.

12. Calculate the number of molecules given the formula and weight of material.

13. Write very large and very small numbers in scientific notation.

Chapter 4 Contents

Part I

Introduction   63

4-1   How Are Molecules of Elements and Compounds Different?   63

4-2   How Are Atoms Held Together?   64

4-3   How Do the Atoms of Elements Share Electrons?   66

4-4   How Do Different Atoms Share Electrons?   68

4-5   How Do Atoms of Metals Combine?   69

4-6   How Does Metallic Bonding Explain Metallic Properties?   70

4-7   What Is a Formula?   71

4-8   How Do Atoms Combine?   72

4-9   What Is a Radical?   73

4-10   How Can Our Valence Table Be Used to Determine
          Chemical Formulas?   74

4-11   How Can We Write Formulas that Contain Radicals?   75

Written Exercises   78

Part II

4-12   How Are Formula Weights Calculated?   85

4-13   What Is the Law of Definite Proportions?   86

4-14   How Can We Determine the Empirical Formula?   88

4-15   How Can We Count Molecules?   90

4-16   How Do Scientists Write Very Large and Very Small Numbers?   91

Chapter 4
How Do Atoms Combine?

Part I

Introduction

In the previous chapter, we learned that atoms of the same element have the same number of protons (atomic number) in their nuclei, and the same electron arrangements. Atoms with different atomic numbers are new elements, with new electron arrangements. Now we are ready to learn how atoms combine with each other.

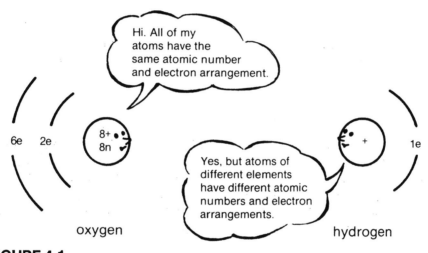

FIGURE 4-1

4-1 How Are Molecules of Elements and Compounds Different?

Our alphabet is made of different letters, just as our universe is made of different elements. When we combine letters in different ways we make words. We can also combine the atoms of elements in different ways to make different **molecules**. Matter is made of particles called molecules. All molecules are made of atoms. When all the atoms of a molecule are alike, with the same atomic number, we have an **element**. A molecule of oxygen is made of two atoms that are held together by a strong binding force. If the atoms of a molecule are not alike, with different atomic numbers, we have a **compound**. A molecule of water is made of two hydrogen atoms that are strongly held to one oxygen atom.

FIGURE 4-2

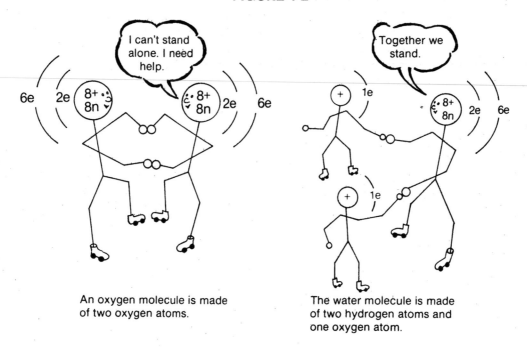

An oxygen molecule is made of two oxygen atoms.

The water molecule is made of two hydrogen atoms and one oxygen atom.

4-2 How Are Atoms Held Together?

The electrons that are farthest from the nucleus are called **valence electrons**. Metals are elements whose atomic nuclei have a weak hold on their valence electrons. Atoms of metals easily lose their valence electrons. The atomic nuclei of non-metal elements have a very strong attraction for their valence electrons. They attract other electrons. When metallic atoms contact nonmetallic atoms, the weakly held valence electrons of the metal may be lost to the nonmetal atom.

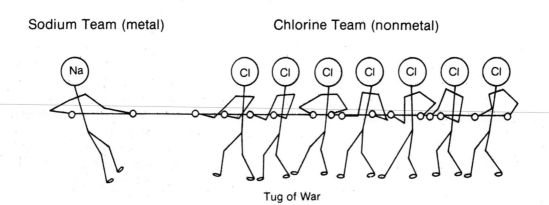

Sodium Team (metal)   Chlorine Team (nonmetal)

Tug of War

FIGURE 4-3

The metal sodium has one valence electron. Chlorine has seven valence electrons. Who will win this tug of war?

After the metallic atoms lose their valence electrons, they are left with too few electrons (fewer electrons in orbit than there are protons in the nucleus). In Chapter 3 we saw that objects with too few electrons are positively charged. A charged atom is called an **ion**. Metallic atoms become positive ions when they lose electrons. Objects with too many electrons are negatively charged. Nonmetallic atoms become negative ions when they gain electrons because then they have too many electrons (more electrons in orbit than there are protons in the nucleus).

FIGURE 4-4

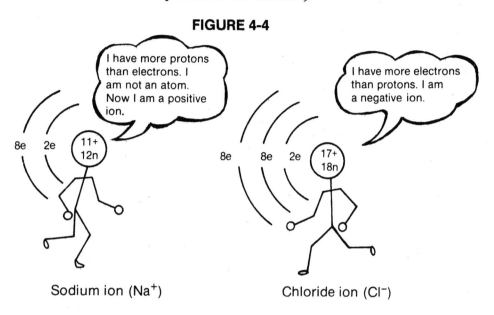

Positive and negative ions attract each other and hold together with great force. This connection between oppositely charged ions is called an **ionic bond**. Ionic materials are formed between active metals and active nonmetals. Common table salt (NaCl) is an example of an ionic material.

FIGURE 4-5

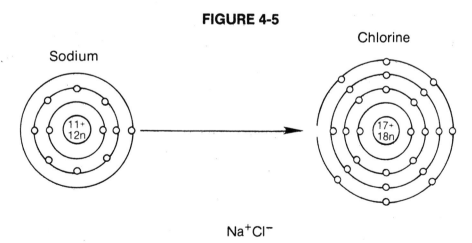

The valence electron of the sodium atom is transferred to the chlorine atom to form sodium chloride.

An ionic bond is an example of what is called a **polar bond**. Polar means "ends with opposite charges on them." Ionic materials like salt (NaCl) are made of positive metallic ions (sodium ions) and negative nonmetallic ions (chloride ions). The positive ion is on one end and the negative ion is on the other, or opposite, end of the bond.

FIGURE 4-6

A sodium chloride crystal shows an array of positive and negative ions.

There are positive ions at one pole and negative ions on the other. It is easy to find the location of the oppositely charged ions in a polar material.

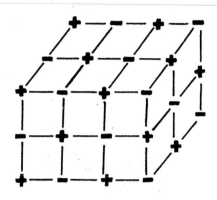

4-3   How Do the Atoms of Elements Share Electrons?

We know that all elements are made of molecules. Nonmetallic molecules, made of two or more atoms of the same element, have an equal attraction for electrons. Neither atom will give up its valence electrons to another atom of the same kind. How can they combine?

FIGURE 4-7

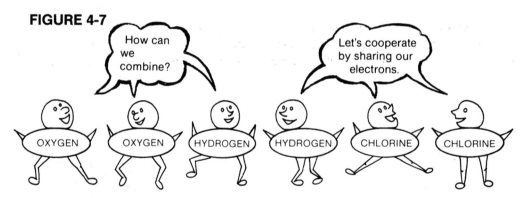

The atoms of nonmetallic molecules bond to each other by sharing each other's valence electrons.

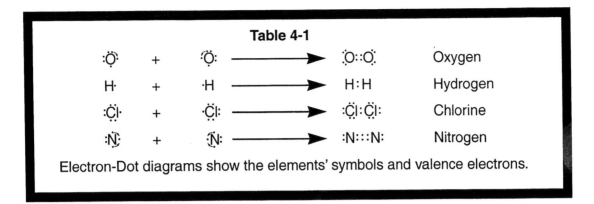

Table 4-1

Electron-Dot diagrams show the elements' symbols and valence electrons.

When atoms of the same element are held together by equally sharing their valence electrons, we have a **nonpolar covalent bond**. The bond is nonpolar because there are no different or opposite ends to the bond. Both ends of the bond have the same electrical charge. The bond is **covalent** because the atoms are sharing each other's electrons.

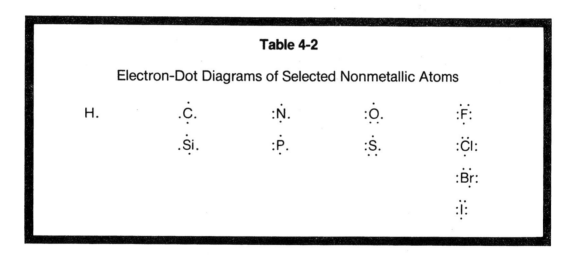

Table 4-2

Electron-Dot Diagrams of Selected Nonmetallic Atoms

Table 4-3

Electron-Dot Diagrams of Covalent Compounds

Using the table of electron-dot diagrams of nonmetallic atoms, draw the electron-dot formulas for the following covalently-bonded compounds.

1. hydrogen sulfide H_2S

2. methyl chloride CH_3Cl

3. hydrogen iodide HI

4. ammonia NH_3

5. methane (natural gas) CH_4

4-4 How Do Different Atoms Share Electrons?

Atoms of the same element share their valence electrons equally because they all have the same attraction for each other's electrons. The atoms of different elements may have a stronger attraction for electrons than other atoms have. They cannot share their electrons equally.

FIGURE 4-8

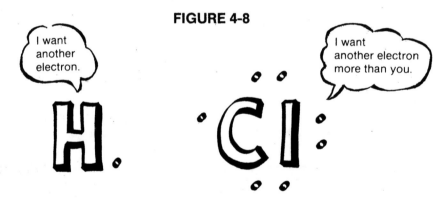

A **polar covalent bond** is formed when two atoms share their electrons **unequally**. The atom with the stronger attraction for electrons draws the electron pair closer to itself and becomes slightly negative as a result. The atom with the weaker attraction for electrons becomes slightly positive because it is more distant from the electron pair. The bond is polar because we can identify a positive end and a negative end to the bond. The bond is still covalent because electrons are still being shared.

FIGURE 4-9

Hydrogen chloride is an example of a material that is polar covalently bonded.

Positive End +

Negative End –

Hydrogen chloride is polar covalently bonded because the hydrogen and the chlorine share a pair of electrons *un*equally. Chlorine has a stronger attraction for electrons than hydrogen has. The electron pair is drawn closer to the chlorine and the chloride end of the molecule is slightly negative, as compared with the hydrogen end of the molecule.

4-5 How Do Atoms of Metals Combine?

We know that metallic atoms have a way of combining with each other. We see metal atoms combined in the form of wire, sheets, and other metallic objects. How are these atoms held together? Metallic atoms all have weak attractions for their valence electrons. When metallic atoms come together, they easily give away their loosely held electrons to neighboring metallic atoms. These electrons move easily between the atoms. As a result, the positive metallic ions are glued together in a sea of mobile electrons. The electrons are distributed throughout the metallic object.

FIGURE 4-10

Metallic Bonding   Hammered or Rolled Metal

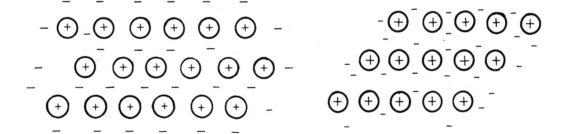

The positive metallic ions are "glued" together in a sea of loosely held electrons.

When metal objects are hammered or rolled into sheets, the ions are squeezed out of place. The metal does not break because the mobile electrons flow with the ions, and they continue to hold the metal together.

4-6 How Does Metallic Bonding Explain Metallic Properties?

When light falls on the surface of a metal, it interacts with the metal's electrons. The electrons act as a force field that stops the light and turns it back. The light is reflected and seen as **metallic luster**. No light can penetrate through any metal. All metals are **opaque**.

FIGURE 4-11

Metals are good conductors of heat and electricity. In Chapter 2, we learned that temperature is a measure of the speed with which particles move. Electrons are always in motion. They vibrate throughout the metallic object. If the temperature is increased on one end of a metal wire or rod, the vibrating motion of the electrons there will speed up. This increased motion will be passed on to its neighboring electrons, thereby conducting heat along the metallic object.

FIGURE 4-12

We have an electric current whenever charges are in motion.

Since the metallic electrons are loosely held, they can easily be pushed from one metallic ion to the next. When a metal is connected to a battery or generator, the electrons are all pushed in the same direction and an electric current is created.

Metals are tough. They are not brittle, nor will they shatter if they are hit with a hammer. When metals are hit, their positive ions are moved but they remain surrounded in their sea of electrons. The strong attraction of the positive ions to their surrounding electrons holds metals together.

Ionic crystals (like salt) are brittle. They are held together by the attraction of positive and negative ions. When they are struck with a hammer, their positive ions may be forced to move next to other positive ions. Negative ions may also be forced next to other negative ions. When this happens, the repulsive forces of the ions will cause the crystal to shatter.

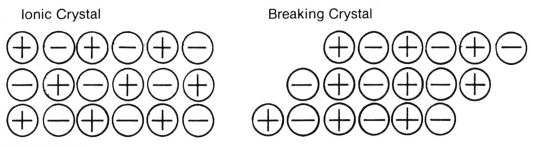

Ionic Crystal   Breaking Crystal

FIGURE 4-13

Ionic crystals are held together by the strong attractive forces of their oppositely charged ions, but these crystals are brittle. If the crystal is hammered, the ions wil be forced to move. Like charged ions will approach each other. They will repel and the crystal will fly apart.

4-7   What Is a Formula?

A chemical formula tells us how many atoms of each element are in the molecule of any substance. The formula for water is H_2O. The "H" is the symbol for hydrogen. Hydrogen is a part of the water molecule. The "O" means that oxygen is also part of the water molecule. The "2" after the H means that two atoms of hydrogen are combined with one atom of oxygen in each water molecule. There is no number after the O. When no number appears after a symbol, it means that there is only one atom of that element. The water molecule is made of two atoms of hydrogen combined with one atom of oxygen. Formulas identify the elements, and the number of atoms, that compose the molecule.

Write the name of each element, and the number of atoms, in the following formulas.

CO_2	H_2SO_4	SiO_2
$NaNO_3$	K_2CO_3	LiF
$AlPO_4$	$MgCl_2$	CaO

4-8   How Do Atoms Combine?

When metals react, their atoms give away their valence electrons. Metallic atom may release one, two, or three valence electrons.

Table 4-4

Electrons in the Valence Shell	Kind of Element	Action
1e	metal	gives up one electron
2e	metal	gives up two electrons
3e	metal	gives up three electrons
5e	nonmetal	accepts three electrons
6e	nonmetal	accepts two electrons
7e	nonmetal	accepts one electron

When nonmetals combine, their atoms accept valence electrons from the metal atoms. They may take one, two, or three electrons. Nonmetallic atoms fill their valence shells to eight electrons. Metallic atoms empty their valence shells when they combine with nonmetals. The **valence** of an element is the number of electrons its atom gives, takes, or shares. The common valences of some elements are listed in Table 4-5.

4-9   What Is a Radical?

Some covalently-bonded groups of atoms act like single atoms. They have a valence and name of their own. These units, or **radicals**, stay together and behave as if they were one new element. Another name for a radical is **polyatomic ion** (many-atomed ion). *Poly-* is a Greek prefix that means "many." Most radicals are negative ions. One common positive radical is the ammonium radical $(NH_4)^+$. The names, formulas, and valences of some common radicals can be found in the "Table of Valences."

Table 4-5: Table of Valences					
METALS					
+ 1		+ 2		+ 3	
Hydrogen	H	Beryllium	Be	Boron	B
Lithium	Li	Magnesium	Mg	Aluminum	Al
Sodium	Na	Calcium	Ca	Iron(III) Fe (Ferric)	
Potassium	K	Strontium	Sr		
Copper (I)	Cu	Iron (II) Fe (Ferrous)			
Mercury (I)	Hg	Mercury (II) (Mercuric)			
Ammonium	NH_4	Copper (II) (Cupric)			
NONMETALS					
– 1		– 2		– 3	
Fluoride	F	Oxide	O	Phosphate PO_4	
Chloride	Cl	Sulfide	S		
Bromide	Br	Sulfate	SO_4		
Iodide	I	Carbonate	CO_3		
Hydroxide	OH	Chromate	CrO_4		
Nitrate	NO_3				
Acetate	$C_2H_3O_2$				
Bicarbonate	HCO_3				
Bisulfate	HSO_4				

Note: The metals have positive valences and the nonmetals have negative valences. Why?

4-10   How Can Our Valence Table Be Used to Determine Chemical Formulas?

We know that the formula for water is H_2O. According to our valence table, hydrogen gives away one electron. Oxygen accepts two electrons. The electron-dot symbols for these elements are H. and :O. These elements combine as shown below.

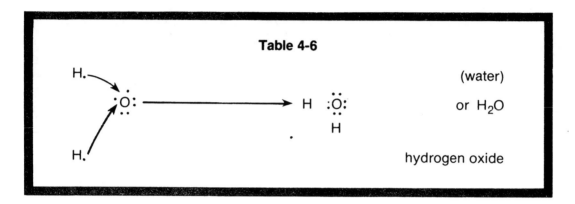

When aluminum combines with oxygen, the compound formed is more complicated.

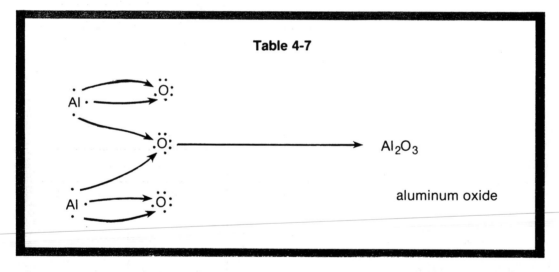

Two aluminum atoms combine with three oxygen atoms to form aluminum oxide. The formula for this compound is Al_2O_3. Determining formulas from diagrams is inconvenient. There is an easier way to write formulas. Let us apply the following rules to the formulas for carbon dioxide and aluminum oxide.

Table 4-8

Rule	Application			
1. Write the symbol of the elements.	C	O	Al	O
2. Always write the metal first, the nonmetal second.				
3. Write the valence of each element above each symbol.	C^4	O^2	Al^3	O^2
4. Crisscross the valences as shown.	C^4 O^2 C_2 O_2		Al^3 O^2	
5. Reduce the subscripts to their lowest terms.	CO_2		Al_2O_3	

4-11   How Can We Write Formulas that Contain Radicals?

Many compounds contain radicals. These formulas can be determined by applying the same rules used for single atoms. Let us apply these rules in determining the formulas for aluminum sulfate and aluminum phosphate. The radicals and their valences can be found in the valence table.

Table 4-9

Rule	Application			
1. Write the symbol of the elements and radicals.	Al	SO_4	Al	PO_4
2. Always write the metal first, the nonmetal second.				
3. Write the valence of each radical or element above each symbol.	Al^3	SO_4^2	Al^3	PO_4^3
4. Crisscross the valences as shown.	Al^3 SO_4^2		Al^3 PO_4^3	
5. Reduce the subscripts to their lowest terms. (If there is more than one radical, brackets must be used.)	$Al_2(SO_4)_3$		$Al_3(PO_4)_3$ $AlPO_4$	

Now You Know

1. Nonmetallic atoms of the same element combine with each other by equally sharing their valence electrons.

2. When atoms share their electrons equally we have **nonpolar covalent bonding**.

3. Nonpolar covalent bonding takes place between atoms that have strong and equal attractions for electrons. It cannot take place between metallic atoms because they have very weak attractions for electrons.

4. Covalent bonding can take place only with unpaired valence electrons of nonmetallic atoms.

5. A single covalent bond is formed by the pairing of one set of electrons.

6. A double covalent bond is formed by the pairing of two sets of electrons.

7. A triple covalent bond is formed by the pairing of three sets of electrons.

8. All chemical reactions between atoms involve the exchange or sharing of valence electrons.

9. The formula identifies the number of atoms of each element that composes the molecule of a compound.

10. Metallic atoms can give away 1, 2, or 3 valence electrons, when they react.

11. Nonmetallic atoms accept electrons from metal atoms, to complete their valence shells.

12. A radical, or polyatomic ion, is a group of atoms that is covalently bonded together, and acts like one atom.

New Words

compound	A material composed of different elements.
covalent bond	Bonding by the sharing of electrons.
element	A material whose atoms have the same atomic number.
formula	A combination of symbols that tells how many atoms of each element composes a molecule.
ion	An electrically charged atom.

ionic bond	The force holding oppositely charged ions together.
metal	A material whose atoms easily lose their valence electrons.
metallic bond	The way in which atoms of metals hold together.
metallic luster	The shine from the surface of a metal.
molecule	The smallest unit of a compound.
nonmetal	A material whose atoms accept valence electrons.
nonpolar	All ends with the same electrical charge.
nonpolar co-valent bond	Equal sharing of electrons between atoms.
opaque	Stops light.
polar	Oppositely charged ends.
polyatomic ion	A radical.
radical	A group of atoms that acts as if it were one atom with its own valence.
subscript	The small number written below the symbol of an element in a formula. It tells how many atoms of that element are in each molecule.
valence	The number of electrons an atom gives, takes, or shares when it bonds with other atoms.
valence electrons	The electrons which are farthest from the atomic nucleus and are involved with bonding.

Formula-Writing Drill

Complete the Following Table

	OH^{-1}	SO_4^{-2}	PO_4^{-3}	S^{-2}	Cl^{-1}
Na^{+1}					
Mg^{+}					
Al^{+3}					
$(NH_4)^{+}$					

Reading Power

For each of the following questions, select one answer that seems most correct.

1. The main idea of the introduction is:
   a. Atoms are made of protons and neutrons.
   b. Electrons orbit the nucleus.
   c. This chapter teaches how atoms combine.
   d. Molecules are made of atoms.

2. "Find the facts" in the introduction.
   a. Choose the correct sentence that explains how the atoms of elements differ.
   b. Choose the correct sentence that indicates the topic of chapter 4.

3. The main idea of section 4-1 is:
   a. The universe is made of elements.
   b. Our alphabet is made of letters.
   c. Elements are made of molecules.
   d. Molecules of elements are made of the same kind of atom. Molecules of compounds are made of different kinds of atoms.

4. "Find the facts" in section 4-1.

   a. Choose the correct sentence that shows that matter is made of molecules.
   b. Find the sentence that explains how molecules of compounds differ from molecules of elements.
   c. Find the words that explain the composition of the water molecule.

5. The main idea of section 4-2 is:

   a. Valence electrons are farthest from the atomic nucleus.
   b. Metallic atoms easily lose their electrons.
   c. Some atoms are bonded together by the attraction of oppositely charged ions.
   d. Nonmetals are made of atoms that gain electrons.

6. "Find the facts" in section 4-2.

   a. Choose the correct sentence that defines **ion**.
   b. Name one example of an ionic material.
   c. What is another name for an ionic bond?

7. The main idea of section 4-3 is:

   a. All atoms combine by ionic bonding.
   b. All atoms combine by nonpolar covalent bonding.
   c. Atoms of the same nonmetallic element combine by forming nonpolar covalent bonds.
   d. Different atoms that form compounds have polar bonds.

8. "Find the facts" in section 4-3.

   Locate the definition of a nonpolar covalent bond.

9. The main idea of section 4-4 is:

   a. Different atoms do not have the same attraction for electrons.
   b. A polar covalent bond is formed when atoms share electrons unequally.
   c. A polar covalent bond is formed when atoms share electrons equally.
   d. A polar covalent bond is formed when metallic atoms lose valence electrons to nonmetallic atoms.

10. "Find the facts" in sections 4-4, 4-5, and 4-6.

    a. In which section is the metallic bond described?
    b. Which section explains that no light can penetrate through a metal?
    c. Which section explains the polar covalent bond?
    d. Which section discusses how metals conduct heat?

11. The main idea of section 4-7 is:

    a. The formula for water is H_2O.
    b. The formula for water tells us that each molecule is made of 2 hydrogen atoms and 1 oxygen atom.

    c.   When metals react, their atoms can give away up to 3 electrons.

    d.   All formulas identify the elements in a compound and tell how many atoms of each element make up the molecule.

12.   The main idea of section 4-8 is:

    a.   When metals react they give away electrons.

    b.   When nonmetals react they gain electrons.

    c.   Metals can give away 3 valence electrons.

    d.   When atoms combine, electrons are involved.

13.   The main idea of section 4-9 is:

    a.   A polyatomic ion is another name for a radical.

    b.   A radical is a combination of atoms that acts as one atom.

    c.   Every radical has its own name and valence.

    d.   Metals have positive oxidation states.

14.   The main idea of section 4-10 is:

    a.   The valences determine the number of atoms that combine in a molecule.

    b.   No formula can be written without valence information.

    c.   The valence table provides essential information for formulas.

    d.   The valence table provides a convenience when writing formulas.

15.   The main idea of section 4-11 is:

    a.   Formulas containing radicals are written by following rules similar to those that apply to atoms.

    b.   When more than one radical appears in a formula, that radical must be bracketed.

    c.   When writing formulas, we always write the metals first.

    d.   Subscripts must be canceled wherever possible.

Mind Expanders

Explain each of the following in your own words, in one or two sentences.

1.   How do atoms of different elements differ from each other?

2.   How do the molecules of elements and compounds differ from each other?

3.   How do atoms become ions?

4.   How do metallic atoms acquire a positive charge?

5.   What is the difference between a polar bond and a nonpolar covalent bond?

6.   Why do atoms of the nonmetals combine to form covalent bonds?

7.   Why do atoms with unpaired valence electrons attract each other?

8.   Which electrons are shown in an electron-dot diagram?

9. Why are oxygen molecules double bonded?

10. Why does hydrogen only form single bonds?

11. Why are atoms, with different electron attractions, not able to share electrons equally?

12. Why are metallic atoms unable to form covalent bonds?

13. How does metallic bonding explain how copper can be drawn into a wire?

14. How do metals conduct electricity?

15. Why are nonmetals poor electrical conductors?

Formulas and Valences of Radicals

Write the formulas and valences of the following polyatomic ions.

phosphate	carbonate	hydroxide	ammonium
nitrate	sulfate	cyanide	bicarbonate
	bisulfate	acetate	

Completion Questions

Complete the following statements using the words below. Some words may be used more than once.

a. atomic number	f. metals	j. polar bond
b. attract	g. molecule	k. subscript
c. different	h. nonmetals	l. symbol
d. ions	i. oppositely	m. valence
e. lose		

1. Atoms of the same element have the same _____ , _____ , and _____ .

2. The molecules of a compound are composed of atoms of _____ elements.

3. The _____ is the smallest unit of a compound.

4. Ions with opposite charges will _____ each other.

5. _____ electrons are farthest from the atomic nucleus.

6. When the atoms of an element enter into a reaction they involve _____ electrons.

7. Metals differ from nonmetals in that metallic atoms _____ valence electrons.

8. Active metals and active nonmetals react to form _____ .

9. Another name for the ionic bond is _____ .

10. Ionic materials are held together by the attractive force of _____ charged ions.

11. The formula of a compound shows the _____ of each element in its molecule.

12. The _____ tells how many atoms of each element are in each molecule of a covalent compound.

13. The combining power, or _____ , is the number of electrons shared by an atom.

14. Atoms of _____ show a weak attraction for electrons.

15. Atoms of _____ have a strong attraction for its electrons.

Multiple-Choice

For each statement or question, select the word or expression that best completes the statement or answers the question.

1. Atoms that are bound together by a nonpolar covalent bond have:
   a. equal attraction for electrons.
   b. unequal attraction for electrons.
   c. molecules that are partly metallic and partly nonmetallic.
   d. oppositely charged ions.

2. An element that has the symbol F is:

    a. a nonmetal that shares one electron.
    b. a nonmetal that shares two electrons.
    c. a nonmetal that shares three electrons.
    d. a metal.

3. Which of these symbols shows a nonpolar covalent bond?

    a. H : B̈r :   b. :C̈l : C̈l:

    c. Na⁺ :C̈l:⁻   d. :N̈e:

4. All of the following are nonpolar covalent bonds except:

    a. H : H   b. :Ö : : Ö:

    c. : N : : : N :   d. H : Ö :
                                   H

5. In order for atoms to form triple bonds, each atom must have:

    a. a single unpaired valence electron.
    b. at least two unpaired valence electrons.
    c. at least three unpaired valence electrons.
    d. a weak attraction for electrons.

Matching

Write the number in column B that relates to the words in column A in the space indicated.

COLUMN A	COLUMN B
_____ nonpolar covalent bond	1. the symbol of the element and its valence electrons.
_____ nonpolar	2. sharing of electrons.
_____ covalent	3. the same ends.
_____ electron-dot symbol	4. two nuclei sharing the same electron pair equally.

Complete the Following Table

Name	Formula	Number of Atoms	Number of Ions
ferric chloride			
sodium phosphate			
calcium bicarbonate			
ammonium sulfate			
magnesium hydroxide			

True or False

If the statement is true, mark it *T*. If the statement is false, change the word(s) in italics, using the list below, to make the sentence true.

polar covalent bond	unequally
nonpolar covalent bond	electron sea
different	metals
lack of	nonmetallic

1. When two atoms equally share a pair of electrons, we have a *nonpolar covalent bond.*

2. All nonmetallic atoms have the *same* attraction for electrons.

3. *Metallic atoms* cannot form covalent bonds with themselves because they have too weak an attraction for an electron pair.

4. *Nonmetals* are good conductors of heat and electricity.

5. Materials that are *good conductors of electricity* are also good conductors of heat.

6. Metallic toughness is caused by the *positive metallic ions.*

7. Metals are shiny because of their *weakly held electrons.*

8. When atoms are held together by *polar covalent bonds*, their atoms share their valence electrons equally.

9. *Nonpolar covalent bonds* are formed between atoms with different attractions for electrons.

10. *Nonmetals* have loosely held electrons.

Part II

4-12 How Are Formula Weights Calculated?

The formula weight is the sum of the weights of all the atoms in a formula. It is easily found using the following procedure:

1. Write the formula.

2. Make the table shown below.

3. Find the total weight of each element by multiplying the number of its atoms by its atomic weight.

4. The formula weight equals the sum of the weights of all the atoms in the formula.

Problem: Find the formula weight of sodium chloride. The formula for sodium chloride is NaCl.

Element Symbol	Number of Atoms		Atomic Weight		Total Weight
Na	1	×	23	=	23
Cl	1	×	35	=	35
				formula weight =	58

Problem: Find the formula weight of aluminum sulfate. The formula for aluminum sulfate is $Al_2 (SO_4)_3$.

Element Symbol	Number of Atoms		Atomic Weight		Total Weight
Al	2	×	27	=	54
S	3	×	32	=	96
O	12	×	16	=	192
				formula weight =	342

Problem: Find the formula weights for each of the following.

HCl	KBr	CaO	NaOH
$CaCl_2$	ZnF_2	$FeCl_2$	AlI_3
$Al(OH)_3$	$Ba_3(PO_4)_2$		$Fe_2(CO_3)_3$

4-13   What Is the Law of Definite Proportions?

A compound is composed of different elements. The law of definite proportions states that *the elements in any compound are in definite proportions (ratios, or percentages) by weight.* In this section, you will learn how to determine the percentage or proportion of the elements in their compounds. The procedure is shown in the following three steps.

Problem: Determine the percent oxygen and hydrogen in water.

Step 1: Find the formula for the compound.
The formula for water is H_2O.

Step 2: Determine the molecular weight of the compound.
The molecular weight is the sum of the weights of all the atoms in the molecule. Atomic weights of all elements can be found in the periodic table.

Element (Symbol)	Number of Atoms		Atomic Weight		Total Weight
H	2	×	1	=	2
O	1	×	16	=	16
			molecular weight	=	18

Step 3: Determine the percent of each element.

$$\% \text{ element } E = \frac{\text{weight of E}}{\text{molecular weight}} \times 100$$

$$\% \text{ oxygen} = \frac{\text{weight of oxygen}}{\text{molecular weight}} \times 100$$

$$\% \text{ oxygen} = \frac{16}{18} \times 100$$

$$\% \text{ oxygen} = 89\%$$

$$\% \text{ hydrogen} = \frac{\text{weight of hydrogen}}{\text{molecular weight}} \times 100$$

$$\% \text{ hydrogen} = \frac{2}{18} \times 100$$

$$\% \text{ hydrogen} = 11\%$$

Conclusion: Water is a compound composed, by weight, of:

89% oxygen and
11% hydrogen.

All of the elements total 100% (89% + 11% = 100%).

Problem: Find the percent of each element in the following compounds. The atomic weights can be found in the periodic table.

1. CO_2	2. $KClO_3$	3. $CaCO_3$
4. Na_2CO_3	5. K_2CrO_4	6. NH_3
7. $CuSO_4$	8. $Zn(NO_3)_2$	9. Fe_2O_3
10. $AlPO_4$	11. $CuCl_2$	12. $Ba(OH)_2$

4-14   How Can We Determine the Empirical Formula?

Empirical means experimental or "from experience." An empirical formula is one that is determined by experiment (i.e., experience). An empirical formula reveals the ratios of the atoms of each element composing the compound in its lowest terms.

Problem: Imagine you are the chief chemist on a space expedition to a new planet. In your explorations, you discover a strange new substance. Your analysis shows that it is composed of 52.2% carbon, 13.0% hydrogen, and 34.8% oxygen. How could you determine the empirical formula for this unfamiliar material?

Solution

1. Since the elements in this substance are in definite proportions, the substance must be a compound.

2. Since the total of these percentages is 100%, it is safe to conclude that all of the elements have been accounted for.

$$(52.2\% + 13.0\% + 34.8\% = 100\%)$$

The formula for this compound will take this form:

$$C_xH_yO_z$$

Our task is to determine the values of x, y, and z.

Step 1

Find the atomic weights of each element, using the periodic table.

The atomic weight of carbon is 12.0.
The atomic weight of hydrogen is 1.0.
The atomic weight of oxygen is 16.0.

Step 2

Divide the % of each element by its atomic weight.

% carbon = 52.2% $\dfrac{52.2}{12.0}$ = 4.35

% hydrogen = 13.0% $\dfrac{13.0}{1.0}$ = 13.0

$$\% \text{ oxygen} = 34.8\% \qquad \frac{34.8}{16.0} = 2.18$$

Step 3

Divide each answer by the smallest number and round off to the nearest whole number. This small whole number tells you the number of atoms of each element in the empirical formula.

For carbon $\qquad \dfrac{4.35}{2.18} = 2 \qquad$ (x = 2)
(There will be 2 carbon atoms in the formula.)

For hydrogen $\qquad \dfrac{13.0}{2.18} = 6 \qquad$ (y = 6)
(There will be 6 hydrogen atoms in the formula.)

For oxygen $\qquad \dfrac{2.18}{2.18} = 1 \qquad$ (z = 1)
(There will be 1 oxygen atom in the formula.)

Answer: The empirical formula of this compound is:

$$C_2H_6O$$

(When no number appears in a formula, one [1] is understood.)

Problem: Determine the empirical formulas for each of the following compounds. (Write the metals first.)

1. 77.8% iron and 22.2% oxygen

2. 39.7% sodium and 60.3% chlorine

3. 36.4% calcium and 63.6% chlorine

4. 43.7% phosphorous and 56.3% oxygen

5. 30.4% nitrogen and 69.6% oxygen

6. 85.7% carbon and 14.3% hydrogen

7. 36% aluminum and 64% sulfur

8. 82.4% nitrogen and 17.6% hydrogen

9. 40% carbon, 6.67% hydrogen, and 53.3% oxygen

10. 42% sodium, 18.9% phosphorous, and 39% oxygen

11. 40.2% potassium, 26.8% chromium, and 33.0% oxygen

12. 16.3% sodium, 38.3% manganese, and 45.4% oxygen

13. 34.5% calcium, 24.1% silicon, and 41.4% oxygen

14. 2.0% hydrogen, 32.7% sulfur, and 66.3% oxygen

15. 1.6% hydrogen, 22.2% nitrogen, and 76.2% oxygen

4-15 How Can We Count Molecules?

The molecular weight is important because it enables us to count molecules.

The molecular weight for water is 18. If you were to weigh 18 grams of water you would have 602,000 billion billion molecules. This number was determined by the Italian chemist **Amedeo Avogadro**. 602,000 billion billion is called the **Avogadro number**. Counting molecules by weight is not new. Banks often count large numbers of coins by weighing them.

Rule: *There are 602,000 billion billion (602,000,000,000,000,000,000,000) molecules in every gram molecular weight.* The molecular weight in grams is called a **mole**.

The number of moles of a substance is computed as follows:

$$\frac{\text{weight of the compound (in grams)}}{\text{molecular weight}} = \text{number of moles}$$

Problem: Find the number of molecules in a. 36 grams of water, and b. 9 grams of water.

---

a. $\dfrac{\text{weight of water}}{\text{molecular weight}} = \text{number of moles}$

$\dfrac{36 \text{ grams}}{18} = 2 \text{ moles of water}$

Each mole has 602,000 billion billion molecules. Two moles of water has 2 × 602,000 billion billion, or 1,204,000 billion billion molecules.

b. $\dfrac{9 \text{ grams}}{18}$ = ½ mole of water

Each mole contains 602,000 billion billion molecules. One half of a mole has ½ of 602,000 billion billion—or 301,000 billion billion—molecules of water.

Find the number of molecules in each of the following.

1. 88 grams of carbon dioxide (CO_2)

2. 16 grams of oxygen (O_2)

3. 270 grams of glucose ($C_6H_{12}O_6$)

4. 17 grams of ammonia (NH_3)

5. 23 grams of alcohol (C_2H_5OH)

4-16 How Do Scientists Write Very Large and Very Small Numbers?

Scientists often find it necessary to write extremely large or extremely small numbers. These numbers have many zeros in them; for example, there are 18 zeros in 1 billion billion. It is inconvenient to write numbers with so many zeros. They are difficult to write and read and the zeros are tedious to count. It is easy to make errors or to count incorrectly using such numbers. To avoid such problems, scientists use a simple system for writing this kind of number. The system is called **scientific notation**.

The rules for scientific notation are as follows:

1. Place the decimal point after the first **digit** (number).

2. Multiply the number by a **power** of 10. The power of a number determines how many places that the decimal point must be moved. If the power is positive, then the decimal point is moved to the right (⎯⎯⎯➤). If the power is negative, the decimal point is moved to the left (◄⎯⎯⎯).

For example, the Avogadro number is 602,000 billion billion or 602,000,000,000,-000,000,000,000. This is 602 followed by 21 zeros. The number, written in scientific notation is 6.02×10^{23}. The power of the 10 is +23. This means that the decimal point placed after the first digit (6), must be moved 23 places to the *right*.

The distance between ions in a salt crystal is about 3/100,000 cm or three hundred thousandths of a centimeter.

This very small number can be written in scientific notation as 3×10^{-5}. The minus five means that the decimal point must be moved five places to the left and the number would look like this: 0.00003.

Write the following numbers in scientific notation.

1. One atom of copper weighs 0.000,000,000,000,000,000,000,10 gram. Express this weight in scientific notation.

2. There are about 10,800,000,000,000,000,000,000 atoms of iron in 1 gram.

3. The mass of the earth is 600,000,000,000,000,000,000,000,000 grams.

4. The mass of an electron is 0.000,000,000,000,000,000,000,000,000,009 gram.

5. The speed of light is 300,000,000 meters per second.

6. The earth is 93,000,000 miles from the sun.

7. The wavelength of green light is 0.000,000,005 cm.

8. 18 molecules out of every 100,000,000 dissolve.

9. The world population is about 4,500,000,000.

10. One water molecule out of every 10,000,000 ionizes.

Chapter 5
Of What Are Solutions Made?

Instructional Objectives

After completing this chapter, you will be able to:

1. Compare fresh water with seawater.

2. Define and give one example of asymmetric, colloid, concentrated solution, decant, dilute solution, distillation, filtrate, filtration, miscible, polar molecule, polar solvent, residue, saturated solution, sediment, solubility, solute, solution, solvent, supersaturated solution, suspension, symmetric, unsaturated solution.

3. Explain distillation.

4. Explain the Tyndall effect.

5. Explain how solutes dissolve in solvents.

6. Explain how dissolved impurities lower the freezing point, and raise the boiling point of the solvent.

7. Define and explain the factors controlling the rate of solution and the extent of solution.

8. Explain how to grow crystals.

9. Interpret solubility curves.

10. Compare and calculate molar concentration of solutions.

Chapter 5 Contents

Part I

Introduction   95

5-1   How Is Seawater Different from Fresh Water?   95

5-2   What Is a Solution?   95

5-3   How Can the Solvent Be Separated from the Solute in a Solution?   98

5-4   What Other Kinds of Solutions Are There?   99

5-5   What Are Polar Molecules?   101

5-6   How Do Salt and Sugar Dissolve in Water?   102

5-7   Why Are Oils Insoluble in Water?   103

5-8   How Does Salt Affect Ice?   103

5-9   How Do Dissolved Solutes Affect the Boiling Point of Water?   104

5-10 How Much Solute Is Enough?   104

5-11 How Much Is Too Much?   106

5-12 How Can We Speed the Rate of Solution?   107

5-13 How Can We Control the Amount of Solute that Can Dissolve?   107

5-14 How Can We Grow Crystals as a Hobby?   109

Written Exercises   112

Graph Interpretation   119

Part II

5-15 Determining the Concentration of Like Solutions   122

5-16 Comparing the Concentration of Different Solutions   123

Molarity Computation Problems   124

Chapter 5
Of What Are Solutions Made?

Part I

> "Water, water everywhere,
> The very boards did shrink;
> Water, water everywhere,
> Nor any drop to drink."

Introduction

These are the words from the "Rime of the Ancient Mariner," written by Samuel Coleridge. It tells of the problem faced by an old sailor dying of thirst on a long ocean voyage. Why can't sailors drink seawater? The world today is suffering from famine in large areas of Asia, Africa, and South and Central America. The cultivated land of the earth cannot produce enough food to feed its human population, and the famine is expected to get worse. If there were enough available fresh water, the earth could be made to yield enough food to feed everyone. Almost eighty out of every one hundred parts of the earth are covered by water—seawater. Why can't farmers use this seawater to irrigate the soil and grow the food that will save millions of lives?

5-1   How Is Seawater Different from Fresh Water?

Seawater has a salty taste. If it is evaporated, you can recover the salts that were dissolved in it. Seawater is a solution of several salts. The amount of salt it contains is greater than your body can accept. The high concentration will dehydrate your body. Seawater also contains dissolved metals that are poisonous. It may also be polluted or contain germs that cause disease. Fresh water is found in lakes, rivers, and reservoirs. Farms can use only fresh water to grow crops. Fresh water, in the form of rain, cleanses the soil by dissolving its salts. As the water becomes salty, it flows to the oceans.

5-2   What Is a Solution?

When you put a spoonful of sugar into a cup of tea, the sugar dissolves. A solution is formed. The sugar is the **solute**. A solute is the substance that is dissolved. The water is the **solvent**. The solvent dissolves the solute. *A solution is a homogeneous (the same throughout) mixture of a solute in a solvent.* The solute particles may consist of atoms, ions, or molecules. A sugar solution is a uniform distribution of molecules of sugar in water. When salt is dissolved in water, we have a distribution of ions. Solutions exist as solids, liquids, and gases.

FIGURE 5-1

Why can't sailors drink seawater?

FIGURE 5-2

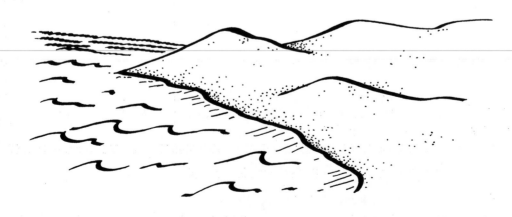

Why can't deserts be changed into farms by seawater irrigation?

FIGURE 5-3a

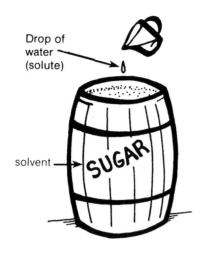

Drop of water (solute)

solvent

SUGAR

FIGURE 5-3b

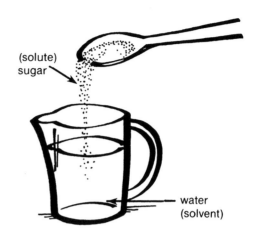

(solute) sugar

water (solvent)

Which is the solute, and which is the solvent?

SOLVENT + SOLUTE = SOLUTION

A solution has more solvent than solute.

Table 5-1   Solutions

Solid Solutions	Liquid Solutions	Gaseous Solutions
alloys	seawater	air
steel (Iron, carbon and other metals)	*tincture of iodine (iodine in alcohol)	
silver amalgam (silver and mercury used for dental fillings)	alcohol in water sugar in water oil in benzene	

*A tincture is a solution in which the solvent is alcohol.

The particle size in a solution is so small that there is no visible interference with the passage of light. The result is that all solutions are clear. Some solutions may have a color, but no solution can be cloudy.

5-3   How Can the Solvent Be Separated from the Solute in a Solution?

The only way to separate the parts of a solution from each other is by the process of **distillation**. First, the solvent is vaporized. Then it is condensed (changed back into liquid) and collected somewhere else. The condensed solvent is called the **distillate**. The remainder consists of concentrated solute.

FIGURE 5-4

Distillation

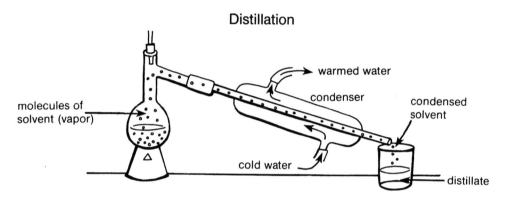

Distillation is expensive because of the cost of the fuel needed to vaporize the solvent. Engineers once attempted to use solar energy to distill seawater to irrigate the deserts. They could not get enough fresh water to make the effort worthwhile. Distillation is a reliable way to remove dissolved poisons from water. Ships at sea, on long ocean voyages, often rely on distilled water for drinking purposes. Solutions administered in hospitals use distilled water for their solvents.

FIGURE 5-5

Solar Distillation of Seawater

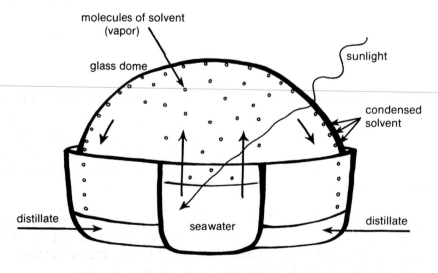

Why can sunlight not solve the world's water problem?

5-4 What Other Kinds of Solutions Are There?

A mixture of clay in water contains suspended particles that are much larger than atoms, ions, or molecules. These particles can be seen and they are large enough to block the path of light. This mixture is called a **coarse suspension**. This type of suspension will be cloudy. The suspended material can be separated from its medium by **filtration** or **decantation**.

FIGURE 5-6   Decantation of wine   **FIGURE 5-7**   Filtration of muddy water

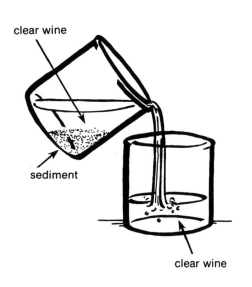

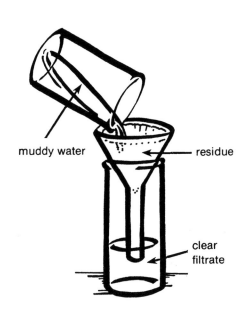

Decantation means that the heavy sediment is allowed to settle, and the clear liquid is carefully poured off.

Filtration is a process in which suspended material is separated from the clear liquid by pouring through a filter. The liquid passes through the holes in the filter, but the suspended particles are too large and remain on the filter.

One of the most common types of solutions is neither a true solution nor a coarse suspension. Many solutions are **colloidal suspensions**. They are dispersions in which the particles are larger than atoms, ions, or molecules, but smaller than the particles in a coarse suspension. Colloids are clear and look like true solutions. The particles are too small to be seen or to block the path of light, but they are large enough to scatter light. The test for a colloid (**Tyndall test**) is to shine a beam of light through a solution.

If the light beam is visible, a colloid exists. The atmosphere is a dispersion of dust in air. You can see the Tyndall effect whenever a sunbeam in the air becomes visible. The light is reflected off the invisible dust particles. Colloids cannot be filtered because particles will pass through the pores in the filter.

Table 5-2

Colloid	Example
solid dispersed in gas	dust in air
solid in a liquid	soap or starch in water
gas in a liquid	whipped cream
liquid in a gas	fog (water in air)
emulsions	milk, mayonnaise, catsup
	ice cream

An emulsion is a suspension of oil in water. All emulsions are white in color. Catsup is red because of the tomato color added to it. Milk is a dispersion of butterfat in water. In order to disperse oil in water, an emulsifying agent must be used. The molecules of these agents are partly soluble in oil and in water. Egg yolk is used as the emulsifier in mayonnaise. Soap is effective as a cleanser because of its ability to emulsify oil in water. In order to be digested, fats that are eaten must be emulsified. The emulsifying agent in your intestine is **bile juice**. The fluid in a blister, as well as many other body fluids, are colloids. Another name for liquid colloids is **sol**. Solid colloids are called **gels**.

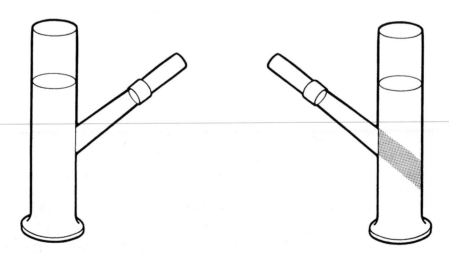

FIGURE 5-8

A true solution is always clear. It will not show the Tyndall effect. Is your drinking water a colloid?

A colloid shows the Tyndall effect. The solution appears cloudy as the suspended particles scatter the light.

5-5   What Are Polar Molecules?

In Chapter 4, we learned that polar means "oppositely charged ends." Many molecules are positively charged on one end, and negatively charged on another end. Water, sugar, and alcohol are examples of substances made of polar molecules.

FIGURE 5-9

Water molecules are asymmetric.

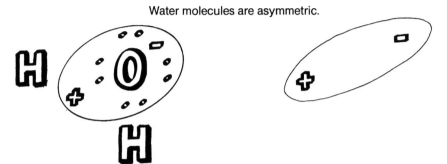

Asymmetric molecules are polar.

FIGURE 5-10

Methane molecules are symmetric.

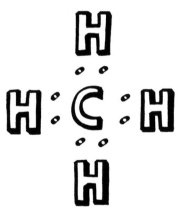

Symmetric molecules are nonpolar.

FIGURE 5-11

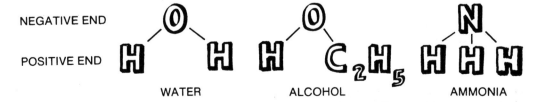

The negative end of polar molecules have strong electron attractors (nonmetals).

In Figure 5-11, it can be seen that polar molecules are not symmetrical (the same on all sides). The nonmetallic atoms have a strong attraction for electrons. In these molecules, the nonmetallic side (oxygen and nitrogen) concentrates the electrons to itself. The nonmetallic side of the molecule becomes negative. The positive end has a deficiency of electrons. Nonpolar molecules do not have opposite ends. Their electrons are evenly distributed at all parts of the molecule.

FIGURE 5-12

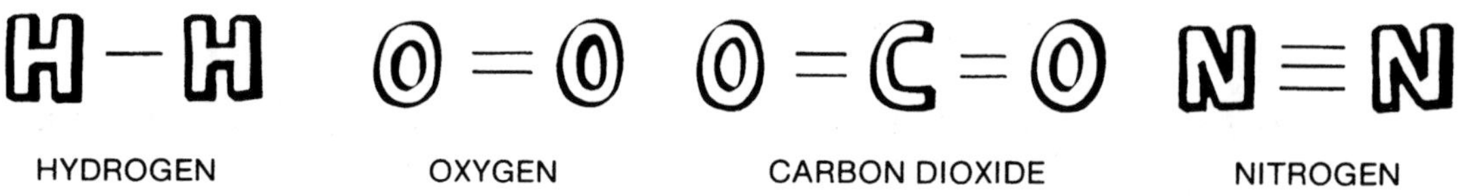

$$H-H \qquad O=O \qquad O=C=O \qquad N\equiv N$$

HYDROGEN   OXYGEN   CARBON DIOXIDE   NITROGEN

Nonpolar molecules do not have different ends. Their electrons are evenly distributed to all parts of the molecule.

5-6   How Do Salt and Sugar Dissolve in Water?

In Chapter 4, section 4-2, we learned that common table salt (NaCl) is an ionic material. When salt dissolves in water, the negative (oxygen) end of the water molecule is attracted to the positive sodium ion. The positive (hydrogen) ends of the water molecule are attracted to the negative chloride ion. The attraction of the polar water molecules to these ions causes the ions to separate from each other. The result is that the ions become dispersed throughout the solution.

Sugar is not ionic, but it is composed of polar molecules. The positive end of the water molecule is attracted to the negative (oxygen) end of the sugar molecule. This results in a dispersion of sugar molecules in water.

FIGURE 5-13

Salt dissolves in water by having its ions separate. A suspension of ions results.

5-7 Why Are Oils Insoluble in Water?

Oil does not mix with water, even though both substances are liquids. The molecules of oil are nonpolar. They are weakly attracted to each other. Polar water molecules are strongly attracted to each other. When oil is added to water, there is no tendency for the oil molecules to break up the strong attraction the water molecules have for each other. As a result, the oil molecules stay by themselves. The water molecules also remain undisturbed. Water molecules are packed more tightly together because of their strong attraction for each other. It is for this reason that water is denser (grams/milliliter) than oil, and oil floats on water.

Oil (liquid fat) is composed of nonpolar molecules. When margarine or butter are stored in your refrigerator, they absorb odors. This happens because odor molecules are also nonpolar.* Nonpolar solutes are soluble in nonpolar solvents.

Solubility Rule: Like Solvents Dissolve Like Solutes.

Salt water   Benzene and corn oil

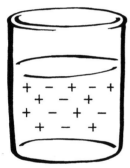  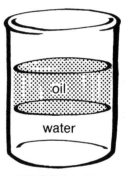

FIGURE 5-14

Ionic salts form a dispersion of ions.

Polar solvents dissolve polar solutes.

FIGURE 5-15

Oil molecules mix well because there are no strong attractive forces to overcome.

Nonpolar solvents dissolve nonpolar solutes.

FIGURE 5-16

Nonpolar oil molecules cannot overcome the strong attractive forces between water molecules.

Oil and water do not mix.

5-8 How Does Salt Affect Ice?

As the temperature falls, water molecules move more slowly, and attractive forces are better able to draw the molecules together. When they lose enough energy, these molecules lock into a rigid crystalline structure—ice. If salt, or any other solute, is dissolved in the water, the solute particles will get in the way of the water molecules as they try to freeze (crystallize). The increased difficulty for water to

*The "old car" odor that develops in automobiles after the first two years or more of use is the result of nonpolar odor molecules, like those of gasoline and oil dissolving in the nonpolar molecular materials that line the inside of the auto. The seat covers, floor mats, and plastic lining of the walls and roof of the automobile are all composed of nonpolar molecular materials. They are impervious to polar solvents like water.

solidify results in a lower freezing point. *The freezing point of all liquids is lowered as a result of dissolved impurities.* It is for this reason that the sanitation department salts the streets in preparation for a severe snowstorm.

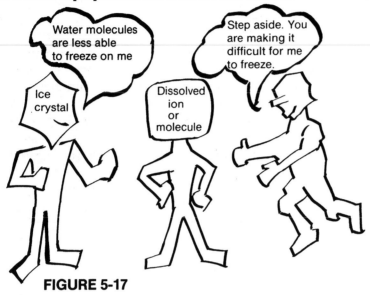

FIGURE 5-17

5-9 How Do Dissolved Solutes Affect the Boiling Point of Water?

When water boils, it is rapidly changed from a liquid into a gas or vapor. Bubbles of vapor will form at the hot bottom of the pot or pan. Boiling is easily recognized when these bubbles reach the surface and burst. In order for the polar water molecules to break free of the strong attraction they have for each other, much energy is needed. Warmer water has more molecules with enough energy to break out of the liquid state than cooler water has. If sugar, salt, or other solutes are dissolved in water, their ions or molecules interfere with the escape of the water molecules from its surface. The result is that a higher temperature is required to reach the boiling point. *The boiling point of all liquids is raised as a result of dissolved impurities.* It is for this reason that motorists add ethylene glycol (antifreeze) to their automobile radiators in the summer, as well as the winter. The antifreeze lowers the temperature at which the water will freeze on a cold winter night, and also raises the temperature at which the water will boil on a hot summer day.

5-10 How Much Solute Is Enough?

Air is a mixture of gaseous molecules. There is no limit to the amount of oxygen, nitrogen, carbon dioxide, or other gases that can be mixed into air. Gases are **miscible** in all proportions. Miscible means "able to be mixed." Liquid solutions of alcohol and water are also miscible in all proportions. There is no limit to how much alcohol can be mixed with water.

FIGURE 5-18

When water boils, bubbles of vapor form at the bottom of the hot container.

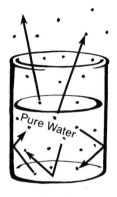

FIGURE 5-19

Water molecules that have enough energy, moving to the surface, can escape the liquid state. Pure water will have a normal boiling point: 100° C.

FIGURE 5-20

Dissolved ions and sugar molecules cannot vaporize. They retard the escape of the liquid and raise the boiling point.

When a little sugar is added to water, it dissolves. If more sugar is added, it will also dissolve. As long as the water is able to continue to dissolve the sugar, the solution is **unsaturated**. When the water has reached its limit, and can no longer dissolve any more sugar, the solution has become **saturated**. A saturated solution is one in which the solvent has dissolved all the solute it can at a given temperature. If more solute (sugar) were to be added to the solution, it would not dissolve. The excess solute would settle to the bottom of its container.

Most solids increase in solubility when the temperature is increased. If a saturated solution is warmed, it may become unsaturated at the higher temperature. An unsaturated solution can be cooled until it becomes saturated.

FIGURE 5-21
Unsaturated Sugar Solution

Unsaturated solutions can still dissolve more solute.

FIGURE 5-22
Saturated Sugar Solution

Saturated solutions have excess solute at the bottom of their containers.

5-11   How Much Is Too Much?

If you hold a jar of honey in your refrigerator too long, the sugar crystallizes, and the honey will not pour. Honey is a **supersaturated** sugar solution. *A supersaturated solution is one in which the solvent is holding more solute than it can normally hold at a given temperature.* Many, but not all solutes, can be made into supersaturated solutions. If a saturated solution of "hypo" (sodium thiosulfate: $Na_2S_2O_3$) is heated, more of its solute will dissolve. Eventually, it will become saturated at the higher temperature. If this warm solution is allowed to cool without any disturbance, a surprising thing takes place. The solution continues to hold its solute at the cooler temperature. The excess hypo remains dissolved in the solution! Since the water (solvent) is now holding more hypo (solute) than it can normally hold when it is saturated at the cooler temperature, the solution is now supersaturated. When this supersaturated solution is disturbed by adding a fresh crystal of hypo to the container, the excess hypo crystallizes out of the solution. A saturated solution results. The heat needed to dissolve this excess hypo is released, and the solution feels warm as a result. Supersaturated solutions are unstable. The excess solute can crystallize out of the solution at any time. Soda is an example of a supersaturated solution of carbon dioxide gas in water.

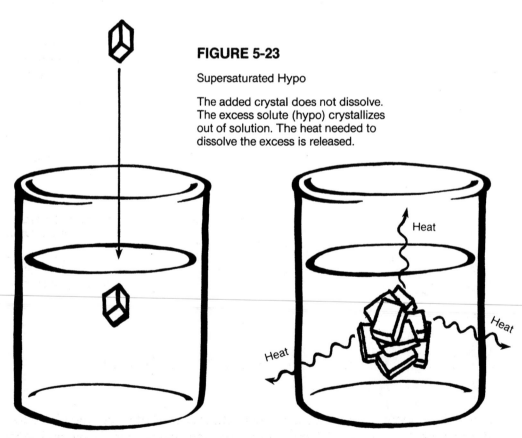

FIGURE 5-23

Supersaturated Hypo

The added crystal does not dissolve. The excess solute (hypo) crystallizes out of solution. The heat needed to dissolve the excess is released.

5-12   How Can We Speed the Rate of Solution?

When you add sugar to your coffee or tea, the sugar dissolves where its surface touches the water. The more contact there is between the sugar and water, the faster the sugar will dissolve. There are three ways to increase surface contact.

a.   A cube of sugar touches the water on six surfaces or sides. If the cube is broken, there will be two more fresh surfaces from which the sugar can dissolve. The broken sugar particles dissolve faster because there is more contact between the solute and the solvent. Powdered or granulated sugar will dissolve faster than large lumps or crystals of sugar.

b.   Another way to increase the contact between solute and solvent is by stirring or agitation. In this way, more fresh water (solvent) is brought into contact with the sugar (solute).

c.   The third way to increase solute-solvent contact is to increase the temperature of the solution. At higher temperatures, all molecules move faster (see section 2-5). The solid solute tends to break down into the liquid state. The solvent also circulates faster. The result is that there is more solute-solvent contact at the higher temperatures, and the solute dissolves faster into the solvent.

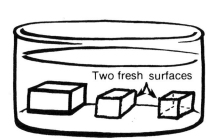

FIGURE 5-24

Smaller particles dissolve faster than large lumps.

Solutes dissolve from their surfaces. Large pieces have less surface contact with the solvent than smaller pieces.

FIGURE 5-25

Stirring speeds solution.

Stirring causes more contact between solute and fresh solvent.

FIGURE 5-26

Solutes dissolve faster at higher temperatures.

Solute and solvent particles move faster at higher temperatures.

5-13   How Can We Control the Amount of Solute that Can Dissolve?

Many clothes washing machines launder with warm or hot water, then have their final rinse with cold water. At higher temperatures, molecules move faster, and solids tend to melt. As a rule, we find that more solid will dissolve at the warmer temperature. When many solute particles are dissolved, the solution is **concentrated**. A **dilute** solution has few solute particles suspended in it. Soaps and detergents are more effective at warmer temperatures, in dissolving dirt and grime. Cold-water detergents are used with fabrics that may be damaged in hot water. The rule is that *solids increase their solubility at higher temperatures.* There are exceptions to this rule. In

FIGURE 5-27

More solid dissolves at higher temperatures.

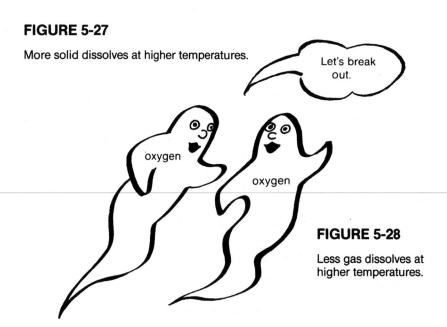

FIGURE 5-28

Less gas dissolves at higher temperatures.

general, calcium compounds are more soluble at cooler temperatures. Washing machines are designed to make the last rinse in cold water to wash out the undissolved calcium salts.

Since molecular motion is increased at higher temperatures, dissolved gases tend to escape from their solutions. The rule for gases is opposite to the rule for solids. *Gases are LESS soluble at higher temperatures.*

5-14   How Can We Grow Crystals as a Hobby?

Crystals exist in different sizes, shapes, and colors. Large and well-shaped crystals can be very valuable and can be sold for much money. Big, beautiful crystals can be set into rings or hung as pendants in pieces of jewelry. Keep in mind that crystals grown from a water solution and worn as jewelry will be damaged in the rain or by moisture. Crystal growing is fun, and it makes use of the chemistry that you have learned. Crystals can be grown at home as a hobby. The materials used are safe when used properly. It is suggested that you start your adventure with alum or blue chrome alum because its crystals will grow rapidly and easily. With experience, you can then go on to grow many other salt crystals. You are reminded, of course, to wash your hands after handling chemicals and before eating to avoid ingesting (eating) chemicals with your food at mealtime.

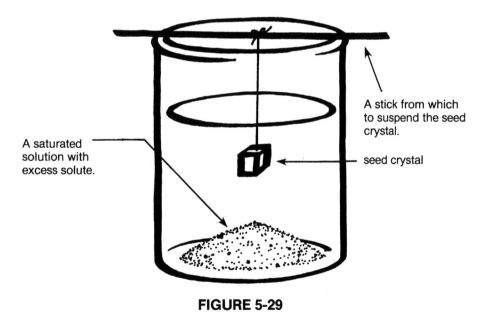

A stick from which to suspend the seed crystal.

seed crystal

A saturated solution with excess solute.

FIGURE 5-29

Crystal Growing at Home

Slow growth makes more perfect shapes.

Now You Know

1. Eighty percent of the earth is covered by seawater, but this water cannot be used to drink or to grow crops.

2. A solution is a homogeneous (the same throughout, or uniform) mixture of a solute in a solvent.

3. The solvent can be separated from its solution by distillation.

4. Distillation is the process through which the solvent is vaporized, then condensed somewhere else in the pure state.

5. A suspension is a mixture in which the suspended particles are larger than atoms, ions, or molecules.

    a. The particles are large enough to be seen.
    b. The particles block the path of light.
    c. Suspensions are not clear. They are cloudy.

6. A colloid is a solution in which the dispersed particles are larger than atoms, ions, or molecules, but still too small to be seen.

7. The parts of a suspension may be separated by filtration or decantation.

8. Polar molecules have oppositely charged ends: a positive end and a negative end. Polar molecules are asymmetrical.

9. Nonpolar molecules do not have oppositely charged ends. They are symmetrical.

10. Polar solutes dissolve in polar solvents.

11. Nonpolar solutes dissolve in nonpolar solvents.

12. Dissolved impurities lower the freezing point and raise the boiling point of a solvent.

13. Gaseous solutions are miscible in all proportions.

14. Some liquid solutions that are miscible in all proportions are alcohol and water, and oil mixtures.

15. When substances are miscible in all proportions, there is no possibility of saturation.

16. A saturated solution is one in which the solvent has dissolved all the solute that it can hold at that temperature.

17. A saturated solution is characterized by excess solute on the bottom of its container.

18. An unsaturated solution is one in which additional solute can be dissolved in the solvent.

19. Most solids increase in solubility with increasing temperature.

20. Gases increase in solubility with decreasing temperature.

21. A saturated solution can become unsaturated if its temperature is increased.

22. An unsaturated solution can become saturated if its temperature is decreased.

23. A supersaturated solution is one in which the solvent is holding more solute than it can hold at saturation at a given temperature.

24. Solutes will dissolve faster if there is more contact between the solute and the solvent. Increased contact can be brought about by increasing the temperature, decreasing the particle size, and by stirring.

25. A concentrated solution has a large amount of solute dissolved in the solution.

26. A dilute solution has a small amount of solute dissolved in its solution.

27. Crystals can be grown at home, or for a science fair project in school. Crystal growing can be an enjoyable hobby.

New Words

asymmetric	Not the same on all sides or ends. Polar molecules are asymmetric.
colloid	A solution in which the solute particles are too small to be seen, but are larger than ions, atoms, or molecules.
concentrated solution	A solution in which a large amount of solute is dissolved.
decant	The process of separating a suspension by allowing the suspended material to settle out, then pouring off the clear liquid.
dilute solution	A solution that has a small amount of dissolved solute.
distillation	The process by which the solvent is separated from the solution.
filtrate	The clear liquid that passes through the holes of a filter.

filtration	The process of separating the parts of a suspension by holding back the suspended material, while the clear liquid passes through small pores or holes of a filter.
miscible	Able to be mixed or dissolved.
polar molecule	A molecule with positive and negative ends.
polar solvent	A solvent whose molecules are polar.
residue	The material held back by a filter.
saturated solution	A solution in which no more solute can dissolve.
sediment	The material that settles out of a suspension.
solubility	The amount of solute that can dissolve.
solute	The part of a solution that is present in a lesser amount. It is the material that is dissolved.
solution	A homogeneous (uniform) mixture of a solute in a solvent.
solvent	The part that dissolves the solute. The solvent forms the largest part of the solution.
supersaturated solution	A solution in which the solvent is holding more solute than it can hold at saturation.
suspension	A mixture of particles that are large enough to be seen.
symmetric	The same on all sides. Nonpolar molecules are symmetric.
unsaturated solution	A solution that can still dissolve additional solute.

Reading Power

For each of the following questions, select one answer that seems most correct.

1. The main idea of the introduction is:
   a. There is not enough fresh water in the world today.
   b. Sailors cannot drink seawater.
   c. Most of the earth is covered with seawater.
   d. Farms cannot be irrigated with seawater.

2. The main idea of section 5-1 is:

   a. Fresh water falls to the earth in the form of rain.
   b. Seawater is a salt solution, and fresh water has very little salt dissolved in it.
   c. Fresh water is found in reservoirs, rivers, and lakes.
   d. Seawater is poisonous.

3. The main idea of section 5-2 is:

   a. Solutions exist as solids, liquids, and gases.
   b. There is less solute than solvent.
   c. All solutions are clear.
   d. A solution is a dispersion of solute in a solvent.

4. The main idea of section 5-3 is:

   a. The solvent may be separated from the solution by distillation.
   b. Distillation is an expensive process.
   c. Distilled water is safe to drink.
   d. The distillate is made of condensed solvent vapors.

5. The main idea of section 5-4 is:

   a. All suspensions are cloudy.
   b. It is easier to separate the parts of a suspension than those of a solution.
   c. In a coarse suspension, the particles are big enough to be seen.
   d. There are other types of mixtures besides true solutions.

6. The main idea of section 5-5 is:

   a. All molecules are polar.
   b. Polar molecules are symmetrical.
   c. Polar molecules have oppositely charged ends.
   d. Water molecules are polar.

7. The main idea of section 5-6 is:

   a. Polar solutes dissolve in polar solvents.
   b. Sugar molecules are polar.
   c. Salt is an ionic substance.
   d. Ionic substances dissociate in water.

8. The main idea of section 5-7 is:

   a. Molecules of oil are nonpolar.
   b. Nonpolar solutes do not dissolve in polar solvents.
   c. Oil floats on water because it is lighter than water.
   d. Oil molecules stick together.

9. The main idea of section 5-8 is:

    a. Salt dissolves in water.
    b. Salt dissolves in ice.
    c. Salt dissolves in both liquid water and ice.
    d. Salt lowers the freezing point of water.

10. The main idea of section 5-9 is:

    a. Any dissolved impurity will raise the boiling point of any liquid.
    b. Dissolved solutes will raise the boiling point of water.
    c. Only sugar and salt can raise the boiling point of water.
    d. Automobile antifreeze is helpful in both winter and summer.

11. The main idea of section 5-10 is:

    a. There is a limit to how much solid or gaseous solute can be dissolved in a liquid solvent.
    b. There is no limit to how much oxygen can be mixed in air.
    c. There is no limit to how much alcohol can be mixed in water.
    d. There is no limit to how much benzene can be mixed in oil.

12. The main idea of section 5-11 is:

    a. Supersaturated solutions are unstable.
    b. Soda is a supersaturated solution of carbon dioxide in water.
    c. Sodium thiosulfate (hypo) solutions can be made supersaturated.
    d. Supersaturated solutions have more solute in solution than they normally have at saturation.

13. The main idea of section 5-12 is:

    a. There are three factors that affect the speed with which solutes dissolve: temperature, particle size, and stirring.
    b. More solid dissolves at higher temperatures, as a rule.
    c. To speed solution, we must increase the contact between solute and solvent. This can be done by applying the factors listed in choice "a."
    d. Stirring is as effective as powdering the solute.

14. The main idea of section 5-13 is:

    a. Concentrated solutions have much solute in solution.
    b. Dilute solutions have little solute in solution.
    c. Solids increase in solubility at higher temperatures.
    d. The solubility of a solute changes with temperature.

15. The main idea of section 5-14 is:

    a. Crystals of value can be grown if care is taken.
    b. Crystal growing can be an enjoyable hobby that is inexpensive.
    c. You should not expose crystals grown from water solutions to moisture.
    d. Crystals grown slowly will be able to grow larger with a more perfect shape.

Mind Expanders

Write your answers to the following questions in two or three sentences.

1. Why can't the oceans be used to solve the world's water shortage?

2. How did the oceans become salty?

3. Why are all solutions clear?

4. How could solar energy be used to distill water?

5. What is the difference between the solute, the solvent, and the solution?

6. Why are some paintbrushes washed with water while other brushes are washed with paint thinner (benzine, turpentine, etc.)?

7. How is the filtrate different from the distillate?

8. Why are symmetrical molecules nonpolar?

9. Why does warm water evaporate faster than cool water?

10. How do polar solutes dissolve in polar solvents?

11. How do dissolved impurities raise the boiling point of water?

12. Why is salt added to ice in a home ice cream maker?

13. How does salt hold water in your body?

14. Why are good cleansers also good emulsifiers?

15. Why are boils soaked in warm salt water? (*Hint:* Colloidal particles have electrical charges on them.)

16. Why do air bubbles form on the sides of a container when tap water is heated?

17. How can a solution be both concentrated and unsaturated at the same time?

18. How can a solution be both dilute and saturated at the same time?

19. Explain two ways for a saturated solution of sugar to become unsaturated.

20. Why does a freshly crystallized supersaturated hypo solution feel warm?

True or False

If the statement is true, mark it *T.* If the statement is false, change the word(s) in italics, using the following list, to make the sentence true.

> decreased   seawater
> dilute   solids
> fresh water   solvent

1. People cannot drink *fresh water* because the salt content is excessive.

2. A *solution* consists of solute in a solvent.

3. An example of a *solid solution* is steel.

4. Deserts cannot be farmed because there is a shortage of *seawater*.

5. *Seawater* has such a high salt concentration that it would draw water from your body.

6. There is always more *solute* in a solution than anything else.

7. A *concentrated* solution has little solute dissolved in it.

8. *Gases* increase in solubility with increasing temperature.

9. A *saturated* solution can become unsaturated if its temperature is increased.

10. Some unsaturated solutions can become supersaturated if the temperature is *increased*.

Matching

COLUMN A	COLUMN B
_____ tincture	1. A cloudy mixture that can be cleared by filtration.
_____ symmetric	2. The solid material that remains after decantation.
_____ suspension	3. Sol.
_____ unsaturated	4. The clear fluid that gets through the pores in the filter.
_____ solute	5. Can dissolve more solute.
_____ residue	6. The same on all sides.
_____ solvent	7. A solution whose solvent is alcohol.
_____ distillate	8. The material that is dissolved.

_____ filtrate   9. The condensed solvent.

_____ colloid   10. The material in which the solute is dissolved.

Multiple-Choice

1. When sugar dissolves in water, the dissolved particle is a(n):

    a. atom   b. ion
    c. molecule   d. large enough to be seen

2. Solution A has 50 grams of salt (NaCl) dissolved in 100 grams of water. Solution B has 10 grams of salt (NaCl) dissolved in 100 grams of water.

    a. Solution A has a higher boiling point and a higher freezing point than solution B.
    b. Solution A has a lower boiling point and a higher freezing point than solution B.
    c. Solution A has a lower boiling point and a lower freezing point than solution B.
    d. Solution A has a higher boiling point and a lower freezing point than solution B.

3. A saltwater solution can be separated into its parts by:

    a. filtration   b. decantation
    c. distillation   d. all of the above (a, b, and c)

4. All suspensions are:

    a. clear and may have color   b. cloudy and may have color
    c. clear and colorless   d. cloudy and colorless

5. All colloids are:

    a. clear and may have color   b. cloudy and may have color
    c. clear and colorless   d. cloudy and colorless

6. A polar molecule is:

    a. asymmetric and has different electrical charges at its opposite ends.
    b. asymmetric and has the same electrical charges at its opposite ends.
    c. symmetric and has opposite charges at its opposite ends.
    d. symmetric and has the same charges at its opposite ends.

7. Sugar dissolves in water because its molecules are:

    a. ionic   b. nonpolar   c. polar   d. nonionic

8. Oil does not dissolve in water because its molecules are:

    a. ionic   b. nonionic
    c. polar   d. nonpolar

9. When ionic salts dissociate in water, the ions:
    a. separate and are associated with water molecules.
    b. remain together and attract water molecules.
    c. remain together and are not attracted to water molecules.
    d. separate and are not associated with water molecules.

10. When benzene is mixed with gasoline, the oil molecules are:
    a. strongly attracted to each other due to polarity.
    b. weakly attracted to each other and are polar.
    c. strongly attracted to each other and are not polar.
    d. weakly attracted to each other and are nonpolar.

11. Dissolved impurities will:
    a. raise the boiling point and raise the freezing point.
    b. lower the boiling point and raise the freezing point.
    c. raise the boiling point and lower the freezing point.
    d. lower the boiling point and lower the freezing point.

12. A true solution is always:
    a. clear   b. cloudy
    c. fit to drink   d. colored

13. Water fit to drink:
    a. never contains bacteria   b. may contain bacteria
    c. always contains bacteria   d. is chemically pure

14. Distillation makes use of the process of evaporation and:
    a. filtration   b. aeration
    c. sublimation   d. condensation

15. A crystal of NaCl added to a solution does not dissolve and the solution remains unchanged. The solution is:
    a. dilute   b. saturated
    c. unsaturated   d. supersaturated

16. When salt water is distilled, the salt is found in the:
    a. distillate   b. coolant
    c. condensate   d. residue

17. In a suspension, the particles:
    a. do not settle to the bottom.
    b. are atomic in size.
    c. are large enough to be seen.
    d. cannot be filtered.

18. As the temperature increases, the solubility of most solids:

    a. increases   b. decreases
    c. remains the same   d. disappears

19. Distilled water is used in hospitals to:

    a. launder linens   b. humidify the air
    c. give intravenous feeding   d. wash dishes

20. As the temperature increases, the solubility of most gases:

    a. increases   b. decreases
    c. remains the same   d. is not affected

Math Skills for Math Power
Graph Interpretation

Refer to the "solubility curves" on the following page to determine the answers to the following problems.

1. Find the temperature at which $KClO_3$ has the same solubility as NH_3.

2. Which solid shows a decrease in solubility with increasing temperature?

3. Which solid shows the greatest increase in solubility with increasing temperature?

4. Which solid shows the least increase in solubility with increasing temperature?

5. How much KNO_3 can dissolve in 100 grams of water at 70° C?

6. How many grams of KNO_3 will crystallize from a solution if it is saturated at 70° C and cooled to 50° C?

7. A saturated solution of NH_4Cl at 50° C is warmed to 70° C. How much more NH_4Cl can be dissolved?

8. Which solid is most soluble in water at 10° C?

9. Which solid is least soluble in water at 10° C?

10. How many grams of NH_4Cl can dissolve in 1 liter of water at 90° C? (*Note:* 1 liter weighs 1,000 grams.)

Figure 5-30

Solubility Curves

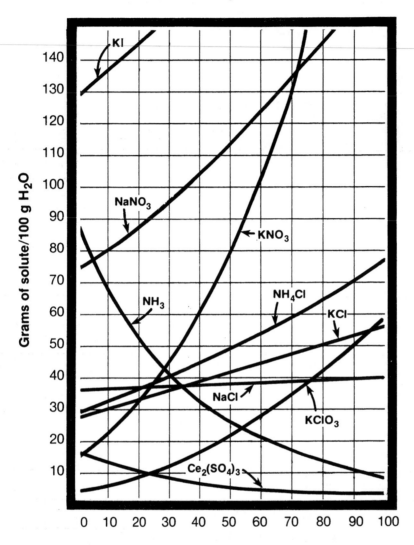

Temperature °C

Figure 5-31

TABLE OF SOLUBILITIES IN WATER											
i—nearly insoluble ss—slightly soluble s—soluble d—decomposes n—not isolated	acetate	bromide	carbonate	chloride	chromate	hydroxide	iodide	nitrate	phosphate	sulfate	sulfide
Aluminum	ss	s	n	s	n	i	s	s	i	s	d
Ammonium	s	s	s	s	s	s	s	s	s	s	s
Barium	s	s	i	s	i	s	s	s	i	i	d
Calcium	s	s	i	s	s	ss	s	s	i	ss	d
Copper II	s	s	i	s	i	i	d	s	i	s	i
Iron II	s	s	i	s	n	i	s	s	i	s	i
Iron III	s	s	n	s	i	i	n	s	i	ss	d
Lead	s	ss	i	ss	i	i	ss	s	i	i	i
Magnesium	s	s	i	s	s	i	s	s	i	s	d
Mercury I	ss	i	i	i	ss	n	i	s	i	ss	i
Mercury II	s	ss	i	s	ss	i	i	s	i	d	i
Potassium	s	s	s	s	s	s	s	s	s	s	s
Silver	ss	i	i	i	ss	n	i	s	i	ss	i
Sodium	s	s	s	s	s	s	s	s	s	s	s
Zinc	s	s	i	s	s	i	s	s	i	s	i

Using the list below, select the term which best complete(s) the statement. You may use the same word(s) more than once.

alloys	saturated
distillation	seawater
increase	solute
miscible	solvent
suspensions	

1. _____ contains too much salt to be useful to farmers.

2. When a cup of water is added to a barrel of salt, the water is the _____ and the salt is the _____ .

3. The _____ becomes the same state of matter as the solvent.

4. All _____ are solid solutions.

5. _____ is the method used to separate the solvent from the solution.

6. Mixtures in which the suspended particles are larger than molecules and able to block the path of light are _____ .

7. A _____ solution is characterized by excess solute on the bottom of its container.

8. As the solvent is allowed to evaporate from its open container, the concentration of the solution will _____ .

9. One way to increase the amount of solid that will dissolve and the speed with which it will dissolve is to _____ the temperature.

10. Alcohol and water are _____ without limit.

Part II

5-15   Determining the Concentration of Like Solutions

If you had two sugar solutions, as shown in Figure 5-32, it should be obvious that solution A is more concentrated than B. In fact, solution A is twice as concentrated as solution B because it has twice as much solute (sugar) dissolved in it. In this case,

FIGURE 5-32

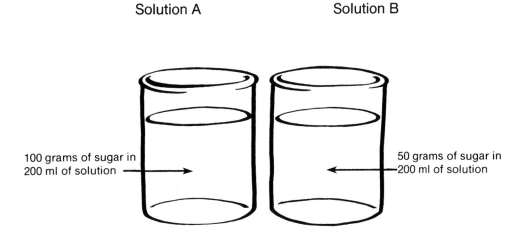

Solution A   Solution B

100 grams of sugar in
200 ml of solution →

← 50 grams of sugar in
200 ml of solution

you could taste each solution. The solution with more suspended sugar molecules would taste sweeter. Of course it is always dangerous to taste laboratory solutions and is a practice that is *never* allowed.

FIGURE 5-33

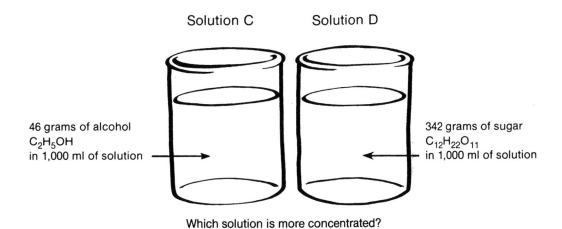

Solution C   Solution D

46 grams of alcohol
C_2H_5OH
in 1,000 ml of solution →

← 342 grams of sugar
$C_{12}H_{22}O_{11}$
in 1,000 ml of solution

Which solution is more concentrated?

5-16   Comparing the Concentration of Different Solutions

We know that more concentrated solutions have more solute molecules or ions in suspension. To determine which solution is more concentrated, we must find a way to count the number of suspended molecules or ions. In section 4-15, we counted molecules by comparing the weight of a substance with its molecular, or formula, weight. In this way, we determined the number of moles we had. Each mole has 6.02 ×

10^{23} molecules. For solutions C and D (Fig. 5-33), we must also compute the number of moles of solute we have in each solution.

Solution C contains 46 grams of alcohol, C_2H_5OH, in one liter of solution. Since the molecular weight for this alcohol is 46, we have one mole of dissolved alcohol in the solution. There are 6.02×10^{23} molecules of alcohol suspended in one liter of solution.

Solution D also has one mole of sugar in solution, because 342 gram equals the molecular weight of the substance. The chemist realizes that both solutions are equally concentrated because they both have equal numbers of suspended molecules in the same amount of solution (one liter).

The chemist expresses the concentration of these solutions in terms of **molarity** (M). Molarity is the number of moles of solute dissolved in each liter of solution.

$$\text{molarity} = \frac{\textbf{number of moles of solute}}{\textbf{liter of solution}}$$

These solutions may be labeled "1 M" to indicate that they are one molar.

Problem: Find the molarity of a solution that contains 23 grams of grain alcohol C_2H_5OH in 1,000 ml of solution (1,000 ml = 1 liter).

1. The molecular weight of C_2H_5OH is 46.

2. The number of moles $= \dfrac{\text{weight of alcohol}}{\text{molecular weight}}$

$$= \frac{23g}{46} = \text{½ mole of alcohol}$$

3. Molarity $= \dfrac{\text{moles of solute (alcohol)}}{\text{liter of solution}}$

Molarity $=$ ½ M or 0.5 M

Problem: Find the molarity of a solution of 684 grams of sucrose (cane sugar), $C_{12}H_{22}O_{11}$, dissolved in 500 milliliters (ml).

1. The molecular weight of sucrose is 342.

2. The number of moles = $\dfrac{\text{weight of sucrose}}{\text{molecular weight}}$

    number of moles = $\dfrac{684 \text{ grams}}{342}$ = 2 moles of sucrose

3. Molarity = $\dfrac{\text{moles of solute}}{\text{liter of solution}}$

    Molarity = $\dfrac{2 \text{ moles}}{0.5 \text{ liter}}$

    (*Note:* 500 ml = 0.500 liter)

    Molarity = 4 M

Problem: Compute the molarity of 46 grams of alcohol dissolved in 200 ml of solution.

1. The molecular weight of alcohol is 46.

2. 46 grams/46 = 1 mole of alcohol

3. Molarity = $\dfrac{1 \text{ mole of alcohol}}{0.2 \text{ liter}}$

    (*Note:* 200 ml = 0.2 liter)

4. Molarity = 5 M

Complete the Following Table

Substance	Molecular Weight	Weight of Substance (grams)	Number of Moles	Volume (ml)	Molarity (m)
NaCl		174		500	
$C_6H_{12}O_6$		90		1,000	
$CaCl_2$		333		250	
H_2SO_4		196		2,000	
K_2CO_3		552		1,500	
NH_4Cl		53		200	
$Al\,(NO_3)_3$		426		100	
NaOH		120		50	
$NaNO_3$		170		250	
HCl		144		500	

Chapter 6

What Are Acids, Bases, and Salts?

Instructional Objectives

After completing this chapter, you will be able to:

1. Recite five examples of acids and bases.
2. List six uses of acids and bases.
3. Identify five substances as acidic or alkaline.
4. Define acid, acidic salt, base, basic salt, hydrogen ion, hydronium ion, indicator, neutral salt, neutralization, pH, strong acid, strong base, weak acid, weak base, salt.
5. Interpret a table of relative strengths of acids and bases.
6. Suggest a test for acids and bases.
7. Determine the acidity of a salt from its formula.
8. Explain the proper first aid for acid or base burns.

Chapter 6 Contents

Part I

Introduction   **129**

6-1   What Is a Strong Acid?   **130**

6-2   What Is a Base?   **132**

6-3   How Can Acids and Bases Be Detected?   **132**

6-4   How Is Acidity Measured?   **133**

6-5   How Do Acids and Bases React with Each Other?   **134**

6-6   Are All Salts Neutral to Litmus?   **135**

Written Exercises   **138**

Part II

6-7   How Is pH Related to Molarity?   **142**

Written Exercises   **145**

Chapter 6

What Are Acids, Bases, and Salts?

Part I

Introduction

Why should batteries be removed from radios, flashlights, electronic calculators, etc., that will not be in use for a long period of time? Why does spoiled milk taste sour? How does overfeeding kill your tropical fish? The answers to these questions can be found by understanding acids.

FIGURE 6-1

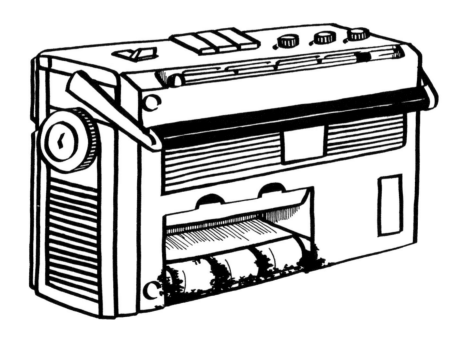

Batteries corrode when left unused in appliances.

Lemons, vinegar, and sauerkraut are similar in that they all have a sour taste. Sauer (sour) is the German word for **acid**. All acids taste sour. Look at the formulas on the following page for each of the following acids. Which element do they all have?

Table 6-1

Chemical Name	Common Name	Formula
hydrochloric acid	muriatic acid	HCl
sulfuric acid	oil of vitriole	H_2SO_4
nitric acid		HNO_3
acetic acid		$HC_2H_3O_2$
carbonic acid		H_2CO_3

**Never taste laboratory acids
or chemicals because they
may be poisonous!**

When acids dissolve in water, they ionize as shown in Table 6-2. Which ion do all acids have in common? In 1887, Swedish chemist Svante Arrhenius defined an acid as a compound that produces hydrogen ions H^+ in water. A hydrogen ion is an atom of hydrogen without its orbital electron. A hydrogen ion is a proton. They are strongly attracted to the negative end of the polar water molecule and form hydronium (H_3O^+) ions. *Acids can be thought of as compounds that donate protons to water molecules to generate hydronium ions.*

6-1   What Is a Strong Acid?

The strength of an acid depends on how much it ionizes. **Hydrochloric acid** is a strong acid in water. It completely ionizes into hydronium and chloride ions. There are no molecules of any strong acid suspended in its solution. Hydrochloric acid, sulfuric acid, and nitric acid are all strong acids. The equation for the ionization of a strong acid, as shown in Table 6-2, has a one-way arrow to indicate that all of its molecules separate into its ions. Weak acids are only partially ionized. Solutions of weak acids contain ions along with molecules of the acid. The equation for the ionization of a weak acid is shown with two arrows. One arrow indicates the breakup of the acid molecule into its ions. The reverse arrow shows the recombination of the ions into its molecule again.

Table 6-2 Comparative Strengths of Acids in Water at 25°C

Name	Ionization Reaction	Strength (How Much It Ionizes)
hydriodic acid	$H_2O + HI \longrightarrow H_3O^+ + I^-$	all ionized (strong acid)
hydrobromic acid	$H_2O + HBr \longrightarrow H_3O^+ + Br^-$	all ionized (strong acid)
hydrochloric acid	$H_2O + HCl \longrightarrow H_3O^+ + Cl^-$	all ionized (strong acid)
nitric acid	$H_2O + HNO_3 \longrightarrow H_3O^+ + NO_3^-$	all ionized (strong acid)
sulfuric acid	$H_2O + H_2SO_4 \longrightarrow H_3O^+ + HSO_4^-$	all ionized (strong acid)
sulfurous acid	$H_2O + H_2SO_3 \rightleftharpoons H_3O^+ + HSO_3^-$	13 molecules ionize out of 100
phosphoric acid	$H_2O + H_3PO_4 \rightleftharpoons H_3O^+ + H_2PO_4^-$	84 molecules ionize out of 1,000
hydrofluoric acid	$H_2O + HF \rightleftharpoons H_3O^+ + F^-$	26 molecules ionize out of 1,000
nitrous acid	$H_2O + HNO_2 \rightleftharpoons H_3O^+ + NO_2^-$	23 molecules ionize out of 1,000
acetic acid	$H_2O + HC_2H_3O_2 \rightleftharpoons H_3O^+ + C_2H_3O_2^-$	4 molecules ionize out of 1,000
carbonic acid	$H_2O + H_2CO_3 \rightleftharpoons H_3O^+ + HCO_3^-$	66 molecules ionize out of 100,000
hydrosulfuric acid	$H_2O + H_2S \rightleftharpoons H_3O^+ + HS^-$	30 molecules ionize out of 100,000
water	$H_2O + H_2O \rightleftharpoons H_3O^+ + OH^-$	1 molecule ionizes out of 10,000,000 (ten million)

6-2 What Is a Base?

Ammonium hydroxide, (NH_4OH), used for cleaning windows, leaves no residue. Lye (sodium hydroxide: $NaOH$) is useful in cleaning clogged drains and pipes. Milk of magnesia (magnesium hydroxide: $Mg(OH)_2$) is an old laxative that is still in use today. Slaked lime (calcium hydroxide: $Ca(OH)_2$) is used by farmers to neutralize acidic soil. All of these substances are **bases**. Look at the formula for each base. Which elements do they all have? All bases taste bitter. Which ion causes this taste? *A base is a compound that produces hydroxide ions in water.*

Table 6-3

Name	Formula	Ionization Reaction
ammonium hydroxide	NH_4OH	$NH_4OH \rightleftharpoons NH_4^+ + OH^-$
sodium hydroxide	$NaOH$	$NaOH \longrightarrow Na^+ + OH^-$
magnesium hydroxide	$Mg(OH)_2$	$Mg(OH)_2 \longrightarrow Mg^{+2} + 2\ OH^-$

The metals lithium, sodium, and potassium (those metals listed in column I of the periodic table) all form strong bases or **alkali**. They are caustic to your skin. This means lye can dissolve or burn your skin and produce serious injury. Strong bases can cause blindness if they get into your eyes. The proper first aid for a chemical burn is to flush the area with cool water, then visit your doctor. There are no molecules of a strong base in water because they are all separated into their ions. The ionization of a weak base is shown in Table 6-3 with reverse arrows. The ions form molecules, at the same time that the molecules form ions. For every 1,000 ammonium hydroxide molecules in water, only four will ionize!

6-3 How Can Acids and Bases Be Detected?

Acids and bases can be detected by their taste, but this practice is dangerous because laboratory chemicals may contain poisons. There are some substances that change their colors between acids and bases. These materials are called **indicators**. **Litmus**, for example, is a dye that is red in the presence of an acid and blue in a base. Other indicators are listed in Table 6-4.

Table 6-4 Acid-Base Indicators

Name	Acidic Color	Basic Color	*pH Range for Color Change
methyl orange	red	orange	3–4.4
congo red	blue	red	3–5
litmus	red	blue	5–8
bromthymol blue	yellow	blue	6–7.5
phenolphthalein	colorless	red	8.3–10
alizarin yellow	yellow	lilac	10–12

*pH is a measure of acidity. See section 6-4.

A common and natural indicator is found in red cabbage. The dye can be extracted from the cabbage by cutting up the cabbage leaves and soaking them in hot water for about one hour. The purple solution that is obtained will turn red in an acid such as acetic acid (vinegar). In a base such as household ammonia, it will first turn bluish and then green. Purple cabbage is grown in a basic soil, and red cabbage is grown in an acidic soil.

6-4   How Is Acidity Measured?

Acidity is a measure of the concentration of hydronium ions in solution. Chemists use a pH scale to express this concentration. A solution with a pH less than 7 is acidic. If a solution has a pH greater than 7, it is basic or alkaline. A solution with a pH of 7 is neutral.

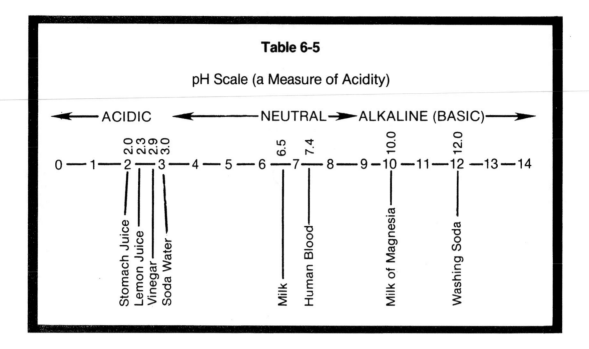

Table 6-5

pH Scale (a Measure of Acidity)

Each whole number difference on the pH scale represents ten times the acidity. A pH of 2 is ten times more acidic than a pH of 3. A pH of 2 is one hundred times more acidic than a pH of 4. The pH of human blood can vary from 7.35 to 7.45. Decaying food often becomes acidic. When milk sours, its pH will decrease. Overfeeding aquarium fish may cause a decrease in the pH because the extra, uneaten food decays, and acids are formed. If the pH falls too low, the fish will die.

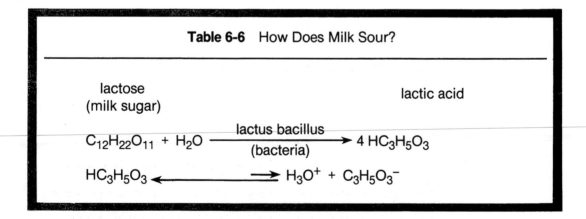

Table 6-6 How Does Milk Sour?

lactose
(milk sugar)   lactic acid

$$C_{12}H_{22}O_{11} + H_2O \xrightarrow[\text{(bacteria)}]{\text{lactus bacillus}} 4\ HC_3H_5O_3$$

$$HC_3H_5O_3 \longleftarrow \longrightarrow H_3O^+ + C_3H_5O_3^-$$

6-5 How Do Acids and Bases React with Each Other?

Sour and bitter are opposite tastes. Acids and bases are also considered opposites, because they react to neutralize each other's properties.

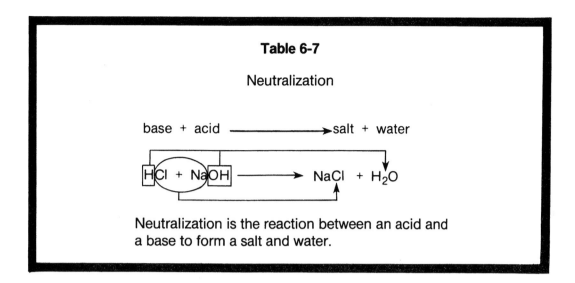

Table 6-7

Neutralization

base + acid ⟶ salt + water

HCl + NaOH ⟶ NaCl + H_2O

Neutralization is the reaction between an acid and
a base to form a salt and water.

When an acid is neutralized by a base, the hydronium ions combine with hydroxide ions to form water. The metal and nonmetal that remain also combine to form a salt. In the reaction between sodium hydroxide (NaOH) and hydrochloric acid (HCl), the salt (NaCl) formed tastes neither sour nor bitter. It has a salty taste. Strong bases have a slippery feeling, but salt water does not feel this way.

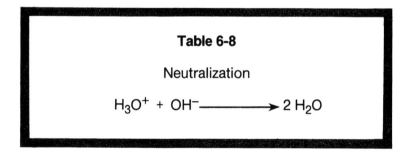

Table 6-8

Neutralization

$$H_3O^+ + OH^- \longrightarrow 2\,H_2O$$

6-6 Are All Salts Neutral to Litmus?

Sodium chloride (NaCl) is a salt formed from a strong acid (HCl) and a strong base (NaOH). The result is that a solution of sodium chloride is neutral. Its water solution has a pH of 7 and the solution has no effect on litmus paper (see Table 6-4). An ammonium chloride (NH_4Cl) solution, on the other hand, will turn litmus red. Ammonium chloride is formed from ammonium hydroxide (NH_4OH), a weak base, and hydrochloric acid (HCl), a strong acid. Salts formed from a strong acid and a weak base are acidic salts. Electric dry cells contain a thick paste of ammonium chloride inside a zinc casing. If these cells were allowed to remain in flashlights and radios for a long period of time, the acidic ammonium chloride would slowly dissolve away the zinc casing and leak out of the cell. This acidic salt would then decay and corrode the metallic parts of your radio or flashlight. Batteries that are not in use should never be stored in any appliances.

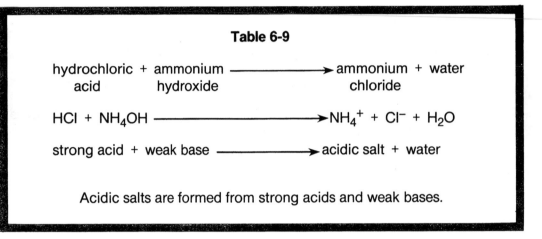

Table 6-9

$$hydrochloric\ acid + ammonium\ hydroxide \longrightarrow ammonium\ chloride + water$$

$$HCl + NH_4OH \longrightarrow NH_4^+ + Cl^- + H_2O$$

$$strong\ acid + weak\ base \longrightarrow acidic\ salt + water$$

Acidic salts are formed from strong acids and weak bases.

Sodium carbonate (Na_2CO_3) is formed from the reaction of a strong base (NaOH) and a weak acid (H_2CO_3). A solution of sodium carbonate turns litmus blue and is basic.

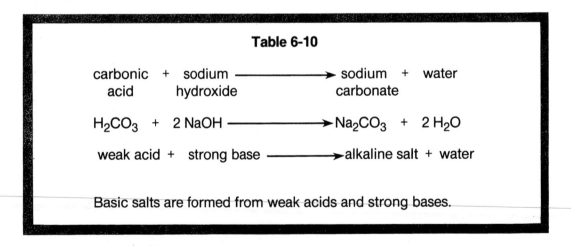

Table 6-10

$$carbonic\ acid + sodium\ hydroxide \longrightarrow sodium\ carbonate + water$$

$$H_2CO_3 + 2\ NaOH \longrightarrow Na_2CO_3 + 2\ H_2O$$

$$weak\ acid + strong\ base \longrightarrow alkaline\ salt + water$$

Basic salts are formed from weak acids and strong bases.

Now You Know

1. Acids are compounds that produce hydronium ions (H_3O^+) in water.

2. Strong acids and bases are completely ionized in water.

3. Weak acids and bases are partly ionized in water.

4. A base or alkaline substance produces hydroxide ions ($OH)^-$ in water.

5. Bases taste bitter. Acids taste sour. Laboratory chemicals should *not* be tasted.

6. The strong acids and strong bases are listed as follows. All other acids and bases may be considered weak.

Table 6-11

Strong Acids		Strong Bases	
hydroiodic acid	HI	lithium hydroxide	LiOH
hydrobromic acid	HBr	sodium hydroxide	NaOH
hydrochloric acid	HCl	potassium hydroxide	KOH
nitric acid	HNO_3		
sulfuric acid	H_2SO_4		

7. The best first aid for chemical burns is to flush the area with water, then visit your doctor.

8. Acids and bases are detected with indicators. Acids turn litmus red and bases turn litmus blue.

9. Acidity is rated in pH units. Acids have a pH range less than 7. Bases have a pH range that is more than 7. Neutral solutions have a pH that is equal to 7.

10. When the pH changes by one whole number, the acidity changes ten times. A pH of 5 is ten times more acidic than a pH of 6.

11. Decaying or rotting food often becomes acidic.

12. Acids and bases neutralize each other to form a salt and water.

New Words

acid	A compound that produces hydrogen ions in water. The hydrogen ions react with the water to make hydronium $(H_3O)^+$ ions.
acidic salt	A salt formed from a strong acid and a weak base.
base	A compound that produces hydroxide ions in water.
basic salt	A salt formed from a strong base and a weak acid.
hydrogen ion	A proton. A hydrogen atom without its electron.
hydronium ion	A proton combined with a water molecule.
indicator	A substance that changes color to indicate acidity.
neutral salt	A salt formed from a strong acid and a strong base.
neutralization	The reaction between an acid and a base to produce a salt and water.
pH	A scale of 0–14 used for measuring the acidity of a solution.
salt	A combination of a metal with a nonmetal.
strong acid	An acid that is completely ionized in water.
strong base	A base that is completely ionized in water.
weak acid	An acid that is partly ionized in water.
weak base	A base that is partly ionized in water.

Reading Power

For each of the following questions, select one answer that seems most correct.

1. The main idea of the introduction is:
   a. Acids are compounds that generate hydronium ions in water.
   b. Lemons, vinegar, and sauerkraut all contain acids.
   c. The hydrogen ion is a proton.
   d. Acids taste sour.

2. The main idea of section 6-1 is:

   a. Strong acids are most dangerous if they touch your skin.
   b. Strong acids are effective when they are concentrated.
   c. Some strong acids are nitric acid, hydrochloric acid, and sulfuric acid.
   d. Strong acids are completely ionized in water.

3. The main idea of section 6-2 is:

   a. Strong bases are completely ionized in water.
   b. Bases taste bitter and have a slippery feel.
   c. Bases produce hydroxide ions in water.
   d. Weak bases are partly ionized in water.

4. The main idea of section 6-3 is:

   a. Litmus is the only indicator used to detect acids and bases.
   b. There are several indicators that can be used to measure the acidity of a solution.
   c. Purple cabbage contains an acid-base indicator.
   d. You should not taste laboratory chemicals.

5. The main idea of section 6-4 is:

   a. The pH of human blood is 7.4.
   b. The pH of a neutral solution is 7.
   c. pH is a measure of the acidity of a solution.
   d. Acidic solutions have a pH of less than 7.

6. The main idea of section 6-5 is:

   a. Acids and bases have many opposite properties.
   b. Acids and bases neutralize each other to produce a salt and water.
   c. A salt is a combination of a metal with a nonmetal.
   d. Water is a combination of a hydrogen ion with a hydroxide ion.

7. The main idea of section 6-6 is:

   a. Acidic salts are formed from a strong acid and a weak base.
   b. Basic salts are formed from a weak acid and a strong base.
   c. Neutral salts are formed from a strong acid and a strong base.
   d. All salts are not neutral to litmus.

Mind Expanders

1. Explain how the storage of old batteries can ruin electrical appliances.

2. Explain how green cabbage can be changed into purple cabbage.

3. Why are there no molecules of nitric acid present in its water solution?

4. Why can't hydrogen ions exist in water?

5. Why is a reverse arrow indicated in the ionization equations of weak acids and weak bases?

6. How do acid-base indicators reveal the acidity of solutions?

7. How can you predict whether a salt is acidic, alkaline, or neutral?

Labeling

Label each compound on the chart as one of the following:

a.	strong acid	e.	acidic salt
b.	strong base	f.	alkaline (basic) salt
c.	weak acid	g.	neutral salt
d.	weak base		

1. $NaOH$	6. $LiNO_3$
2. HNO_3	7. $HC_2H_3O_2$
3. H_3PO_4	8. $NaC_2H_3O_2$
4. K_2CO_3	9. $Al_2(SO_4)_3$
5. $Mg(OH)_2$	10. $ZnSO_4$

pH Table Completion

Table 6-12			
Solution	**Acidic, Alkaline or Neutral**	**Effect on Litmus**	**pH (more, less or equal to 7)**
Na_3PO_4			
H_2O			
LiOH			
HI			
$CuSO_4$			
$Pb(NO_3)_2$			

Completion Statements

Complete the following neutralization reactions.

    a. Fill in the formula for each missing compound.

    b. Label the ACID, BASE, SALT, and WATER.

    c. Determine if the salt is acidic, alkaline, or neutral.

    d. Label each acid and base as strong or weak.

    e. Name each compound.

1. $3\,NaOH + H_3PO_4 \longrightarrow$ _____ $+ 3\,H_2O$

2. $Fe(OH)_3 + 3\,HCl \longrightarrow$ _____ $+ 3\,H_2O$

3. $2\,KOH + H_2SO_4 \longrightarrow$ _____ $+ 2\,H_2O$

4. _____ $+$ _____ $\longrightarrow KNO_3 +$ _____

5. _____ $+$ _____ $\longrightarrow NaCl +$ _____

Multiple-Choice

1. The formula for sulfuric acid is:

   a. HCl b. HNO_3 c. H_2SO_4 d. H_2CO_3

2. A neutral solution will:

   a. have no effect on litmus
   b. turn litmus blue
   c. turn litmus white
   d. turn litmus red

3. The acid found in vinegar is:

   a. sulfuric b. acetic c. hydrochloric d. nitric

4. To test for an acid, use:

   a. blue litmus
   b. a glowing splint
   c. red litmus
   d. limewater

5. The chemical formula for lye is:

   a. $Ca(OH)_2$ b. KOH c. NH_4OH d. NaOH

Part II

6-7 How Is pH Related to Molarity?

Water molecules are polar. They have strong attractions for each other. These attractions are so strong that water is able to ionize itself (although only one molecule out of every 10 million [$1/10,000,000$ or 10^{-7}] will ionize).

$$H_2O \longleftrightarrow H^+ + OH^-$$

For every hydrogen ion formed, a hydroxide radical is also formed.

Pure water is neutral because it ionizes to form equal concentrations of hydrogen and hydroxide ions. In pure water, the molar concentration of hydrogen ion is 10^{-7} M (see section 5-16). The hydroxide ion concentration is also 10^{-7} M.

$(H)^+$ Concentration = $(OH)^-$ Concentration = 10^{-7} M.

The product of these ion concentrations is called the **solubility product** for water. Other substances may have solubility products that are different from those of water. The solubility product (K_w) is constant (the same) at a fixed temperature.

$$K_w = (H)^+ \times OH^-$$

$$K_w = (10^{-7}) \times (10^{-7})$$

$$K_w = 10^{-14}$$

We add powers (exponents) when we multiply.

The chemist finds it more convenient to express the hydrogen ion concentration in terms of **pH**. The "p" is the power (exponent) for the 10 without its minus sign. The "H" represents the hydrogen ion. The hydrogen ion concentration in pure water is 10^{-7} M, and its pH = 7. Since $(H)^+ = (OH)^-$, the $(OH)^- = 10^{-7}$ M. The pOH also equals 7.

In pure water pH = 7 and pOH = 7

pH + pOH = 7 + 7

pH + pOH = 14

Problem: Find the a. $(H)^+$, b. pOH, c. (OH) of a solution whose pH = 3. Identify the solution as acidic, alkaline, or neutral.

a. If the pH = 3, find (H)$^+$.

$(H)^+ = 10^{-3}$ M. (3 is the power of the ten.
We must insert its minus sign.)

b. Find pOH.

$$pH + pOH = 14$$

$$\begin{array}{r} 3 \;\; + pOH = 14 \\ -3 \qquad\quad = -3 \\ \hline pOH = 11 \end{array}$$

c. Find (OH)$^-$.

If pOH = 11 then

$(OH)^- = 10^{-11}$ M

This solution is acidic because the hydrogen ion concentration is greater than the hydroxide ion concentration. $[(H)^+ > (OH)^-]$. *Note:* 10^{-3} is a much greater number (100 million times greater) than 10^{-11}.

Complete the Following Table

Table 6-15				
pH	$(H)^+$ Molarity	pOH	$(OH)^-$ Molarity	Acidic Basic Neutral
3	10^{-3}	11	10^{-11}	acidic
1				
	10^{-12}			
		12		
			10^{-10}	
		1		
			10^{-5}	
	10^{-11}			
		7		
5				
			10^{-4}	
	10^{-8}			
6				

Chapter 7

Energy Forms and Nuclear Chemistry: World Problem or Solution?

Instructional Objectives

After completing this chapter, you will be able to:

1. Define atomic fission, atomic fusion, energy, fossil fuel, fuel, geothermal energy, half-life, nuclear reactor, photosynthesis, radioactivity, solar energy.

2. Identify the source of energy for our planet, cite examples of different energy forms, and trace them to their ultimate source.

3. Explain how the stars and the sun generate energy.

4. List five energy alternatives to the fossil fuels.

5. Discuss strengths and weaknesses associated with each energy source.

6. Name three types of nuclear radiation and give their:

      a. Greek symbol   b. charge   c. mass

7. Explain the term half-life, and solve half-life problems.

8. Explain two ways that radiation can damage cells.

9. Name three types of radiation detectors.

10. Explain how electricity is generated by nuclear reactors.

11. Discuss the advantages and disadvantages of nuclear power plants.

Chapter 7 Contents

Part I   Energy Forms

Introduction **149**

7-1   Where Does Our Energy Come From?   **149**

7-2   How Does the Sun Gets Its Energy?   **149**

7-3   Why Can't We Depend on Fossil Fuels to Meet Our Energy Needs?   **151**

7-4   Can Solar Energy Solve Our Energy Problems?   **152**

7-5   Can Other Sources of Energy Solve Our Energy Problems?   **155**

Part II   Nuclear Chemistry

7-6   How Was Nuclear Radiation Discovered?   **156**

7-7   What Is Radioactivity?   **156**

7-8   How Does Radiactive Decay Change Elements?   **158**

7-9   How Is Radioactivity Detected?   **159**

7-10   How Does Radiation Affect Living Things?   **159**

7-11   How Are Radioisotopes Used?   **160**

7-12   How Is the Rate of Radioactive Decay Measured?   **161**

7-13   What Is Nuclear Fission?   **161**

7-14   How Can a Nuclear Reactor Generate Electricity?   **163**

7-15   Can Nuclear Energy Solve Our Energy Problems?   **164**

Written Exercises   **167**

Career Information   **175**

Chapter 7

Energy Forms and Nuclear Chemistry: World Problem or Solution?

Part I Energy Forms

Introduction

Energy can be thought of as the ability to force an object to move a distance. The most abundant form of energy, which is free and available to everyone, is sunlight. Of course you can't get something for nothing. Engineers know that the cheaper to obtain and more abundant a form of energy is, the harder it is to harness. The more expensive and difficult to obtain a form of energy is, such as coal or oil, the easier it is to use. As our world exhausts its reserves of oil, and oil prices fluctuate, new sources of energy must be found. The availability of energy affects the living standards of every living person and those yet to be born. The hope for our future is in the development of new technology for abundant and inexpensive energy.

7-1 Where Does Our Energy Come From?

The energy needed to maintain life comes from the food we eat. We must have complex carbohydrates because we derive energy from the breakdown of these materials. Green plants absorb water and carbon dioxide. With the help of sunlight, the plant constructs carbohydrate molecules. This process, called **photosynthesis**, is how the energy of the sun is captured and stored for living things.

When you turn on the lights in your home, electricity surges along the wires. This electricity, obtained from your utility company, is produced by generators. These machines are driven by steam that comes from burning coal, gas, and oil. These fuels are called **fossil fuels** because they are derived from the remains of plant and animal life. Fossils are the remains of things that lived a long time ago.

7-2 How Does the Sun Get Its Energy?

Almost all of the energy on our planet is derived from our closest star, the **sun**. All stars, as well as the sun, are natural devices that convert matter into energy. On

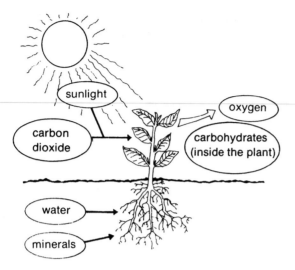

FIGURE 7-1

All energy for living things comes from the sun.

Carbon dioxide + water + sunlight ⟶ carbohydrates

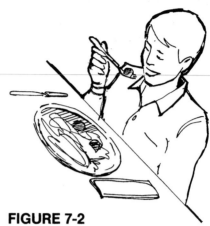

FIGURE 7-2

Food energy is stored sunlight. It is the most expensive form of energy.

FIGURE 7-3

Energy from coal and oil can heat our water and our homes. Coal and oil are forms of stored sunlight.

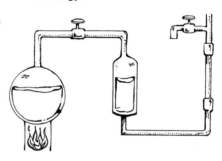

Table 7-1			
Energy Costs for Some Foods and Fuels			
Food or Fuel	**Cost Per Unit**	**Energy per Pound (Kilocalorie)**	**Cost per Million Energy Units**
bread	$ 1.00/lb	1209	$ 819.00
butter	$ 1.35/lb	3614	$ 373.00
sugar	$ 0.89/lb	1864	$ 477.80
sirloin steak	$ 3.58/lb	836	$ 4,280.00
Scotch whiskey	$ 30.00/gal	1173	$ 3,840.00
coal	$ 80.00/ton	3200	$ 12.40
gasoline	$ 1.95/gal	5582	$ 61.50
electricity	$ 0.04/KWH		$ 48.30

Courtesy: U.S. Nuclear Regulatory Commission

the sun, hydrogen isotopes (see section 3-6) are fused together into helium. **Fusion** is the merging of light atomic nuclei into heavier ones. This process results in the release of enormous amounts of energy. The source of energy for the sun and all stars is from **nuclear fusion**.

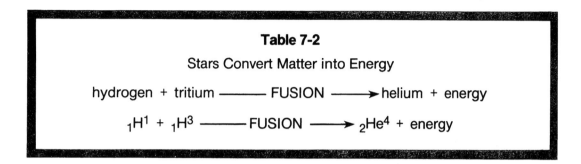

Table 7-2

Stars Convert Matter into Energy

hydrogen + tritium ——— FUSION ——➤ helium + energy

$_1H^1 + _1H^3$ ——— FUSION ——➤ $_2He^4$ + energy

Energy is released during the fusion of hydrogen and tritium into helium. Tritium is the rarest isotope of hydrogen. It has an atomic weight of three.

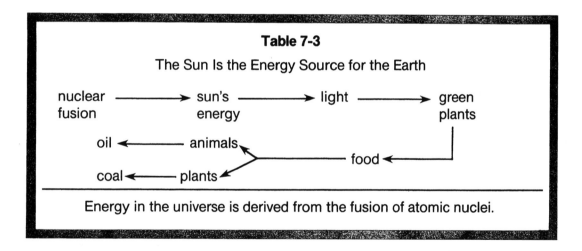

Table 7-3

The Sun Is the Energy Source for the Earth

nuclear ——➤ sun's ——➤ light ——➤ green
fusion   energy   plants

oil ◀——— animals
coal ◀——— plants ——➤ food ◀———

Energy in the universe is derived from the fusion of atomic nuclei.

7-3 Why Can't We Depend on Fossil Fuels to Meet Our Energy Needs?

In section 7-1, it was explained that coal, gas, and oil are called fossil fuels because they are formed from the remains of living things. Oil is an excellent fuel, but there are many problems associated with it. Much oil is imported, creating a dependence on foreign suppliers. It is also subject to price manipulations.

There is an abundance of coal in the United States, but coal is not without its problems. If we were to use coal at the same rate we use oil, the supply would be consumed within the next 150 years. The mining of coal presents problems. Both underground and surface mining are dangerous. There are about six times as many accidental deaths each year in the extraction of coal as there are in the mining of nuclear fuel (uranium). The surface mining of coal results in the destruction of the landscape. It is expensive to restore the land and the lost topsoil. Nature forms only one inch of topsoil every 500 years.

Air pollution is a major cause of premature death. Coal releases more nuclear radiation into the air from smokestacks than any nuclear reactor does. Coal-fired electrical generating plants release particles even when clean-up devices are in operation. These particles are so small they can penetrate deep into the lungs and cause disease. Poisonous gases such as sulfur dioxide cause much lung irritation. Sulfur dioxide released through the chimney, and sulfur exposed in the mining process, are the causes of sulfuric acid (H_2SO_4) rains (see section 13-5d). These **acid rains** are destroying our forests and farmlands. They are killing the life in our rivers and lakes.

The continued burning of coal, gas, and oil causes the release of much carbon dioxide into the air. Scientists believe that an overabundance of carbon dioxide could produce changes in the climate of the earth. If the earth should get warmer, the ice at the north and south poles would melt. The rising sea level could flood all of the world's coastal cities. Manhattan would become the Venice of the United States (see section 13-7a).

7-4 Can Solar Energy Solve Our Energy Problems?

Every one or two weeks, the earth receives the sunlight equal to all the fossil fuel stored on this planet. If we could use all of the sunlight that reaches the United States, we could generate twenty-five times the electrical needs of the country! For every 100 units of solar energy that reaches the earth, 45 units reach the ground, 40 units are reflected back into space, and 15 units are absorbed by the air.

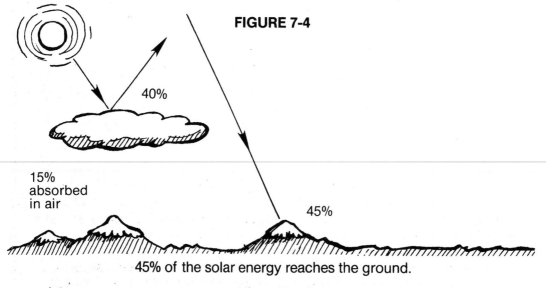

FIGURE 7-4

40%

15%
absorbed
in air

45%

45% of the solar energy reaches the ground.

Solar energy is cheap, inexhaustible, and clean. The sun radiates about the same amount of energy every day. Solar energy is responsible for the winds. The big problem with this energy is finding how to capture and convert the solar energy we receive into a useful form.

FIGURE 7-5

These solar cells are able to change sunlight into electricity.

Courtesy: Socony Mobil Oil Co.

FIGURE 7-6

Small private homes in rural areas heat their water with sunlight.

For every 10,000 units of energy that reach a green plant, about 1,000 units are changed into additional plant material (growth). 9,000 energy units are expended in the other living processes. Plant-eating animals (herbivores) consume the plants. For every 1,000 energy units they take in, they convert only about 100 units into new flesh. Life processes demand the other 900 energy units. When you eat a steak, your body adds about 10 energy units out of every 100 units eaten. Solar energy is spread thinly, and is not used efficiently in nature.

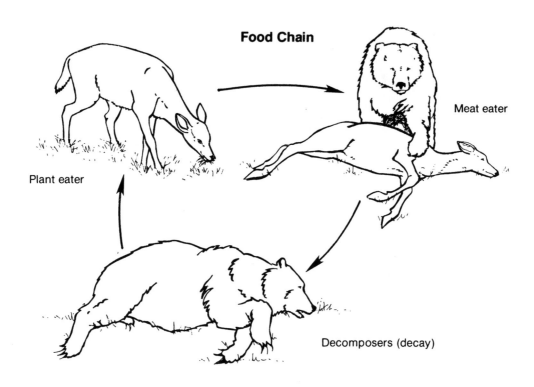

Food Chain

Meat eater

Plant eater

Decomposers (decay)

FIGURE 7-7   All life forms are related to each other.

Table 7-4	
Units of Energy Used for Growth	
sun	10,000
green plants (producers)	1,000
herbivores (plant eaters)	100
carnivores (meat eaters)	10
decomposers (bacteria)	1

Solar energy is not without environmental and economic difficulties. This form of energy must be concentrated. Although sunlight is free, solar devices are expensive. The materials needed to construct solar devices (copper and aluminum) are made available at great energy costs (see sections 9-10 and 9-13). It requires much money and energy to make use of solar energy. While solar energy may be clean, obtaining the materials for solar devices contributes to pollution. Solar energy has an enormous potential to help heat homes and to provide domestic hot water.

FIGURE 7-8

This city apartment building heats its water with sunlight. Notice the solar panels on its roof.

7-5 Can Other Sources of Energy Solve Our Energy Problems?

There are many other ideas for the generation of usable energy. Some engineers suggest using the interior heat of the earth. This is called **geothermal energy**. This technology is useful in areas where water can be pumped down into deep holes in the earth to the hot rocks below. Steam is produced and used to generate electricity.

Geothermal energy, wind power, tide utilization, fermentation to alcohol, and natural gas from decaying garbage have limited usefulness. At present, schemes to change coal into usable liquid and gaseous fuels are too expensive and have many problems associated with them.

Part II   Nuclear Chemistry

7-6   How Was Nuclear Radiation Discovered?

In 1896, Henri Becquerel, a French physicist, placed a salt of uranium on photographic film that was wrapped in two sheets of thick black paper. When the film was developed, he was surprised to discover that the salt had taken a picture of itself! Becquerel concluded that an invisible form of light came from the uranium atoms. This light was able to penetrate through the wrappings of the film. Two chemists, Pierre and Marie Curie, wanted to know the source of this new and strange form of radiation. Their work led them to discover a new metal. They named it "radium" because its atoms radiated this new invisible, powerful, and penetrating light. Marie Curie called the ability of a substance to produce these powerful rays "radioactivity." We know that radiation comes from the breakdown of unstable atomic nuclei.

7-7   What Is Radioactivity?

While the Curies were working in France, Ernest Rutherford was studying for his doctoral degree in England. His teacher was Professor John J. Thompson (section 3-3). He was very excited about the new discoveries of the French chemists. He obtained a small sample of radium and made an extraordinary discovery from a simple, but revealing experiment. Rutherford placed a small sample of radium into a small hole he had drilled in a block of lead. In this way he was able to obtain a narrow radioactive ray or beam. Then he held a magnet near the ray. He knew that an electrically-charged ray would bend in the presence of a magnet, while an uncharged light beam would be unaffected. To everyone's surprise, he obtained three rays: a positive ray, a negative ray, and an uncharged ray. The positive ray was found to consist of helium nuclei (a unit of two protons and two neutrons). The negative ray revealed itself to be a beam of electrons. The uncharged ray was a high-energy X ray. He named the three rays after the first three letters of the Greek alphabet: **alpha** (α), **beta** (β), **gamma** (γ). Becquerel, the Curies, Thompson, and Rutherford were all awarded Nobel prizes for their outstanding achievements.

Courtesy: EPA

FIGURE 7-9a

The cooling towers of a nuclear reactor release water vapor into the air.

FIGURE 7-9b

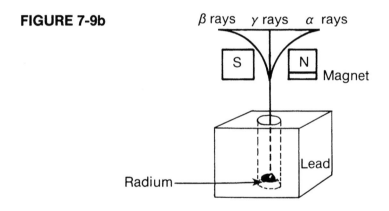

Table 7-5 Radioactive Radiations			
	Charge	**Mass**	**Symbol**
Alpha Ray (helium nucleus)	+ 2	4	$_{+2}He^4$ (α)
Beta Ray (electrons)	– 1	0	$_{-1}e^0$ (β)
Gamma Ray (X ray)	0	0	γ

7-8   How Does Radioactive Decay Change Elements?

All elements whose atomic numbers are greater than 82 are radioactive. They all decay into stable isotopes of lead. In addition, there are many other radioactive isotopes. Today, we know of more than 1,300 radioactive isotopes. Whenever the atomic numbers (number of protons) change, a new element is formed. A radioactive decay reaction that results in a change from one element into another is called a **transmutation reaction**. In the following transmutation equations, the symbol represents the nucleus of the element. The superscript indicates the atomic weight (mass number), and the subscript shows its atomic number or charge.

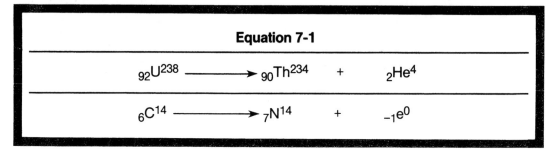

Equation 7-1

$$_{92}U^{238} \longrightarrow {}_{90}Th^{234} + {}_{2}He^{4}$$

$$_{6}C^{14} \longrightarrow {}_{7}N^{14} + {}_{-1}e^{0}$$

When writing nuclear equations, weight (mass) cannot be gained or lost. The sum of the mass numbers of the reactants must be equal to the sum of the mass numbers of the products. Likewise, electrical charge cannot be gained or lost. The algebraic sum of the charges must be equal on both sides of the arrow. Negative electrical charges are indicated with a minus sign. A beta decay (electron emission) results in an increase in the atomic number of the new element.

In nuclear reactions, one unknown element in the equation can be identified by finding its mass and charge.

Equation 7-2

$$_{90}Th^{234} \longrightarrow {}_{n}X^{m} + {}_{2}He^{4}$$

The mass of X can be found as follows:	The charge of X can be found as follows:
m + 4 = 234	n + 2 = 90
m = 234 − 4	n = 90 − 2
m = 230	n = 88

According to the periodic table, element number 88 is radium (Ra) (isotope 226).

7-9   How Is Radioactivity Detected?

Radioactivity is detected by what it does. One method used to detect radioactivity is with **photographic film**. Wrapped film is darkened by exposure to radiation. The amount of radiation is measured by the darkness (density) of the film. X-ray technicians, and others who work with nuclear substances, wear film badges to monitor their exposure to nuclear radiation.

Alpha, beta, and gamma radiations interact with atoms and molecules and knock out their electrons. When this happens, ions are produced. **Ionization detectors** such as the **Geiger-Muller counter**, **bubble chamber**, and **cloud chamber** make use of this phenomenon.

Radiation is also measured with a **scintillation counter**. This device contains a crystal (sodium iodide and thallium iodide) which produces a series of flashes when exposed to radioactive materials. The number of flashes each second measures the amount of radiation present, and the brightness of the flash indicates the energy or intensity of the radiation.

7-10   How Does Radiation Affect Living Things?

Alpha particles are massive and cause the most ionization. Fortunately, alpha particles experience the most difficulty in penetrating any media. They are usually stopped by a centimeter of air, a sheet of paper, or your skin. Beta particles travel more easily through barriers, but they cause less damage and ionization than alpha particles. Gamma radiation is the most penetrating, but produces the least ionization. If these radiations enter the body, they can damage or destroy the cells. It is interesting to note that while radioactive materials can cause cancer, they are also used to treat cancer because they attack cancerous cells.

Radiation is **cumulative** for all living things. That means that radiation builds or adds to all other radiation to which your body is exposed in its lifetime. Unnecessary radiation should be avoided wherever possible. Excessive exposure to sunlight can cause skin cancer.

Table 7-6

Source of Radiation	Millirems/year
Cosmic radiation	50
Earth	47
Building materials	3
Air	5
Elements in the body	21
TOTAL	126
Medical (X rays, therapy, etc.)	74
Nuclear industries, etc.	2
TOTAL (natural + human sources)	202

A **rem** is a measure of absorbed radiation. The recommended dose limit for humans is 500 millirems per year. (1,000 millirems equal 1 rem.)

Table 7-7
Lethal Dose Values for Various Organisms

Organism	Rems	Organism	Rems
human	400	dog	325
bacteria	5,000-13,000	rabbit	300
virus	100,000-200,000	rat	850
guinea pig	200	monkey	450

The lethal dose for humans is 400 rems; i.e., 50% death rate is expected within 30 days of exposure.

7-11 How Are Radioactive Isotopes Used?

Nuclear medicine makes use of radiation and radioisotopes to detect and treat many medical disorders. Radioactive isotopes, which are quickly eliminated from the body, are used for diagnostic injections. Technetium (99) is used for pinpointing brain tumors. Cardiologists use radioactive thallium (201) to watch and photograph the function of the heart as blood surges through it. Iodine (131) is used to treat hyperthyroidism. These isotopes, when allowed to enter the body, identify themselves by their constant radiation. This allows the doctor to trace the "tagged" atoms and follow their chemical reactions.

Radioisotopes are not confined to medical applications. Carbon (14) was used to determine the age of relics of organic origin such as the Dead Sea scrolls. The ratio of uranium (238) to lead (206) in a mineral can be used to determine the age of the mineral.

Radiation kills bacteria, yeasts, molds, and insect eggs in foods, permitting the food to be stored for long periods of time. The CAT scanner is a radiation device that allows the doctor to view cross sections of the body without any surgery. Pictures are taken and disorders confirmed and pinpointed. Nuclear magnetic resonance (NMR) allows the monitoring of nonradioactive isotopes. This offers the advantage of avoiding exposure to radiation, and the new element formed, after the radioactive decay is completed.

The energy derived from radioactive isotopes was used to power communications equipment on space probes around the planet Saturn. This planet receives only 3% of the sunlight that we enjoy on Earth. Solar cells would not be a suitable power source in this remote region.

7-12   How Is the Rate of Radioactive Decay Measured?

The speed with which radioactive atoms decay cannot be controlled. The number of atoms that will decay in a given unit of time is different for each isotope. The time needed for half of its atoms to decay is called the **half-life** of an element. It is determined experimentally. Some half-lives are given in Table 7-8. Isotopes with short half-lives are suitable for medical use.

7-13   What Is Nuclear Fission?

The earth, as it travels through space, moves through a sea of nuclear particles that originates on the sun and distant stars. We call these radiations **cosmic rays**. Among this nuclear debris are neutrons. They reach us on the earth. They penetrate the walls of our homes and burrow into the interior of the earth itself. If a slow moving neutron should strike a uranium (235) nucleus, the nucleus would split. There would be a great release of energy and three more slow-moving neutrons would be produced. This phenomenon is called **atomic fission**. If each neutron produced in the reaction were to cause other uranium atoms to undergo fission, the reaction would accelerate. A cataclysmic explosion would result.

Table 7-8

Isotope	Half-life	Isotope	Half-life
C^{14}	5730 years	P^{32}	14.3 days
Co^{60}	5.3 years	K^{42}	12.4 hours
Cs^{137}	30.23 years	Sr^{90}	28.1 years
Fr^{220}	27.5 seconds	Tl^{201}	74 hours
H^{3}	12.26 years	Ra^{226}	1,600 years
U^{238}	4.51×10^{9} years	I^{131}	8.07 days
K^{40}	1.28×10^{9} years		
U^{235}	7.1×10^{8} years		

FIGURE 7-10 *Nuclear Fission*

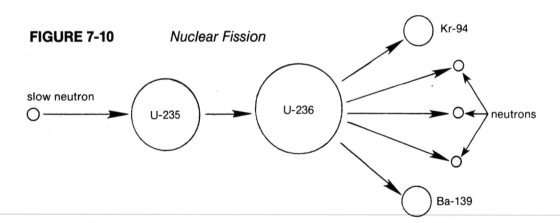

Equation 7-3

$$_{92}U^{235} + {}_{0}n^{1} \longrightarrow {}_{36}Kr^{94} + {}_{56}Ba^{139} + 3{}_{0}n^{1} + \text{energy}$$

Equation 7-3 is a nuclear fission reaction of the original atom bomb that was exploded over the Japanese cities of Nagasaki and Hiroshima. It is also one of the reactions found in our nuclear reactors.

Uranium isotope 235 is fissionable, but it is a rare isotope. Uranium ore contains only 0.7% U-235. Most uranium on our planet is U-238, which does not fission. It is difficult to obtain the conditions for an atomic explosion. A small, slow-moving missile— a neutron—must make a direct hit on a small target, a U-235 nucleus.

In order to make a fission bomb, the target must be increased in size so that the chance for a slow-moving neutron to fission a U-235 nucleus is great. This is accomplished by concentrating the U-235 so that the neutrons from one fission reaction will not escape nearby nuclei. When this happens, the fission process is self-sustaining. A chain reaction takes place. One fission reaction causes three more to occur, and each succeeding fission reaction causes two or three more such reactions. An atomic bomb requires about 20 kg of 97% enriched U-235. This is "weapons grade" uranium.

Fissionable nuclear fuel is rare on earth. To ensure an ample supply of fuel for use in nuclear reactors (and weapons), fuel is produced in a **breeder reactor**.

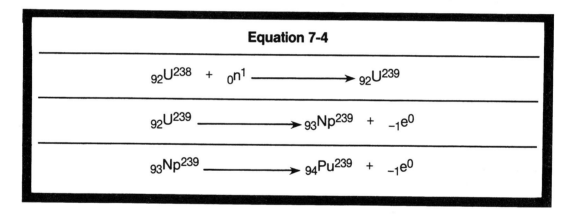

Equation 7-4

$$_{92}U^{238} + \ _{0}n^{1} \longrightarrow \ _{92}U^{239}$$

$$_{92}U^{239} \longrightarrow \ _{93}Np^{239} + \ _{-1}e^{0}$$

$$_{93}Np^{239} \longrightarrow \ _{94}Pu^{239} + \ _{-1}e^{0}$$

The product of the breeder reactor is **plutonium**. It is probably the most potent poison known to humans. If a gaseous compound of plutonium were inhaled, lung cancer would result in due time.

7-14   How Can a Nuclear Reactor Generate Electricity?

Commercial nuclear reactors have 3 to 4% enriched U-235 or Pu-239. The fuel is not concentrated enough to go out of control and blow up as an atom bomb. The fuel mixture is rich enough, however, to produce a sustained chain reaction. For every U-235 or Pu-239 nucleus that fissions, one neutron fissions another nucleus, while the other neutrons escape to freedom. As the atomic nuclei of the fuel undergo fission, energy in the form of heat is released. The heat is used to make steam, which drives generators. The reactor is equipped with neutron-absorbing control rods. They are made of boron or cadmium steel. When the rods are thrust into the reactor, neutrons are absorbed. With fewer neutrons available for fissioning atoms, the reactor is slowed down. If the rods are pulled out, fewer neutrons will be absorbed, and more fission reactions will take place. The reaction will generate more heat for more steam and increase the output of electricity.

There are several different types of nuclear reactors, but they all have certain components. In addition to the fuel and control rods, all nuclear reactors need a moderator, a coolant, and a protective shield. The moderator may consist of graphite (carbon), wax, or heavy water (deuterium oxide). The moderator serves to slow down fast-moving neutrons. This ensures an ample supply of slow-moving neutrons for the chain reaction. The coolant is needed to dissipate the heat produced by the fission reaction. Water is the most common coolant. The reactor's fuel core is surrounded by a containment vessel. It is lined with thick reinforced concrete blocks to protect personnel from radiation.

7-15   Can Nuclear Energy Solve Our Energy Problems?

Though nuclear power reactors cannot explode like nuclear bombs, many other problems are associated with their use. Perhaps the most serious of these is the storage of nuclear waste. Because several fission products are highly radioactive for hundreds or even thousands of years, they must be isolated from the living world for a very long time. Nuclear reactors produce several tons of these materials each year they operate.

Severe problems with radioactive waste at U.S. government nuclear weapons installations in Hanford, Washington, Rocky Flats, Colorado, and Savannah River, South Carolina, will cost hundreds of billions of dollars. These sites may never be cleaned up. They could remain off limits to humans for centuries.

Every power reactor in the country has its own nuclear waste dump. As the power plants run, more wastes build up on site. Because no permanent waste-storage facility is available, each reactor could become a forbidden, contaminated zone, not unlike the weapons facilities mentioned above. The waste problem limits nuclear power as an energy choice.

A long-term possibility for nuclear energy is fusion. In section 7-2 we learned that the atomic nucleus is the source of energy of the sun and stars. This energy is derived from the fusion of protons and neutrons (nuclei of hydrogen isotopes) into helium.

Fusion power reactors are a long way off. In order to run, they must maintain temperatures near that of the sun itself (100 million degrees Celsius). This problem currently cannot be overcome because today's fusion reactors, like the **Tokomak**, cannot produce more energy than is required to maintain the reaction. Nor can they operate for long periods of time.

University of Utah experiments in "cold fusion" at much lower temperatures briefly promised to get around these problems. But those results should be treated skeptically because they have not been repeated by other scientists. Fusion will not solve our energy problems any time soon.

Courtesy: Princeton University Plasma Physics Lab

FIGURE 7-11

Tokomak Fusion Test Reactor (TFTR)　　　Can artificial stars light our homes?

Now You Know

1.　Energy is the ability to force an object to move a distance.

2.　The United States and the entire world are coping with a severe energy problem.

3.　All sources of energy are either difficult to obtain or to use.

4.　Almost all energy on the earth is derived from the sun.

5.　Food, coal, oil, and natural gas are forms of stored sunlight.

6.　The sun derives its energy from the fusion of hydrogen.

7.　The energy obtained from the foods we eat is more costly than the energy obtained from fossil fuels.

8.　Fossil fuels are coal, gas, and oil. They are called fossil fuels because they are derived from the remains of ancient life forms.

9.　Coal is being used more and more to meet our energy needs as oil prices fluctuate.

10. Coal presents several problems as a fuel.

    a. Coal is dirty and causes severe pollution problems as it burns.

    b. Coal mining is dangerous.

    c. Surface mining of coal results in the destruction of the land.

    d. Use of coal produces sulfuric acid rains that kill the life in our rivers, lakes, farmlands, and forests.

    e. Extensive burning of coal, or any carbon fuel, may cause climatic changes for the planet due to excessive amounts of carbon dioxide in the air.

    f. Air pollution from a coal stack releases more nuclear radiation into the air than any nuclear reactor.

    g. A coal stack releases particles that are so small they can penetrate deep into your lungs.

11. Solar energy is free and abundant. It is always there, but it cannot be used to solve our great need for energy.

    a. Solar energy is so thinly diffused that it cannot be sufficiently concentrated.

    b. Solar energy involves hidden costs and pollution.
        (1) The metals needed for solar devices are expensive.
        (2) These metals require much energy to be extracted from the ground.
        (3) Extracting these metals from the raw ore creates much pollution.

12. Transmutation takes place whenever there is a change in the atomic number.

13. Radioactivity is dangerous to your health.

14. Nuclear energy, in the form of the nuclear reactor, can meet our energy needs, but not without problems.

    a. Nuclear reactors are expensive.

    b. There is much concern about the disposal of used nuclear fuel.

    c. There is concern about possible accidents involving nuclear reactors.

15. In the distant future, possibly within thirty years, it is hoped that nuclear fusion can be used to generate an abundance of energy.

New Words

atomic fission	The splitting of heavy atomic nuclei into lighter nuclei, with the release of much energy.
atomic fusion	The merging together of light atomic nuclei (hydrogen) into heavier nuclei, with a great release of energy.
energy	The ability to do work, or the ability to force an object to move a distance.

fossil fuel	A fuel that is formed from the remains of ancient life. Coal, gas, and oil are examples of fossil fuels.
fuel	A material from which energy can be derived.
geothermal energy	The heat from the inside of the earth. This heat is used to make steam for the generation of electricity.
half-life	The time needed for one half of the radioactive atoms in a sample to decay.
nuclear reactor	A controlled nuclear fission device for the generation of energy.
photosynthesis	The process by which a green plant uses sunlight to make carbohydrates.
radioactivity	The emission of energy from a disintegrating atomic nucleus.
solar energy	The energy from the sun (sunlight).
transmutation	The change of one element into another.

Reading Power

For each of the following questions, select one answer that seems most correct.

1. The main idea of the introduction is:
   a. Solar energy is the most abundant form of free energy.
   b. Cheap and easily obtained energy is hard to use.
   c. Expensive and hard to obtain energy is easy to use.
   d. The world is caught in the grip of an energy shortage.

2. The main idea of section 7-1 is:
   a. Carbohydrates are a form of stored sunlight.
   b. Green plants need sunlight in order to make food.
   c. The sun is the source of most energy on the earth.
   d. Coal and oil are forms of stored sunlight.

3. The main idea of section 7-2 is:
   a. The source of solar energy is atomic fusion.
   b. Green plants are responsible for the food supply.
   c. The source of all energy on the earth is the sun.
   d. The sun is our closest star.

4. The main idea of section 7-3 is:

    a. Coal mining is dangerous.
    b. Fossil fuels can no longer meet our energy needs.
    c. Coal burning releases particles into the air.
    d. Coal burning pollutes the air.

5. The main idea of section 7-4 is:

    a. Solar energy cannot solve the energy problem.
    b. Solar energy is too diffuse and needs to be effectively concentrated.
    c. Only 45% of the solar energy that falls on the earth reaches the ground.
    d. Solar energy is free and inexhaustible.

6. The main idea of section 7-5 is:

    a. Geothermal energy looks most promising.
    b. Wind power may be used extensively.
    c. There are many energy sources that are useful.
    d. There are many energy sources, but none are able to meet the energy needs of our society.

7. The main idea of section 7-6 is:

    a. Radioactivity is a natural phenomenon.
    b. Radioactivity is a human invention.
    c. Radioactivity was discovered by its effect on film.
    d. Radioactivity is everywhere.

8. The main idea of section 7-7 is:

    a. Radium was discovered by Marie Curie.
    b. Radioactivity is the release of energy by unstable atomic nuclei.
    c. Radioactivity can be used to power deep space probes.
    d. Radioactivity can be a threat to your health.

9. The main idea of section 7-8 is:

    a. There are many radioactive atoms.
    b. Energy is released from radioactive atoms.
    c. Energy is stored in atomic nuclei.
    d. When the atomic number changes, a new element is formed.

10. The main idea of section 7-9 is:

    a. Radiation is detected by what it does.
    b. Radiation is detected by photographic film.
    c. Radiation is detected by a Geiger counter.
    d. Radiation is detected by a cloud chamber.

11. The main idea of section 7-10 is:

    a. Radiation is used to treat some cancers.
    b. Unnecessary radiation should be avoided.
    c. Radiation destroys living tissue.
    d. Radiation is a useful tool in medicine.

12. The main idea of section 7-11 is:

    a. Radioisotopes are used in medical research.
    b. Radioisotopes are used to power equipment on spaceships.
    c. Radioisotopes are used in industry.
    d. Radioisotopes are used in all of the above.

13. The main idea of section 7-12 is:

    a. Radioactive decay is measured by half-life.
    b. All radioisotopes have a half-life.
    c. Doctors only use isotopes with short half-lives.
    d. Radioactive half-life is uncontrollable.

14. The main idea of section 7-13 is:

    a. Nuclear fission takes place on the sun.
    b. Nuclear fission is the splitting of heavy atomic nuclei.
    c. Nuclear fission is used in atom bombs.
    d. Nuclear fission is used in nuclear reactors.

15. The main idea of section 7-14 is:

    a. Nuclear reactors directly change nuclear energy into electricity.
    b. Nuclear reactors are very expensive to build.
    c. Nuclear reactors generate steam to drive electrical generators.
    d. Reactor waste disposal is a problem.

16. The main idea of section 7-15 is:

    a. Nuclear reactor accidents are feared by people.
    b. Nuclear waste limits the role of nuclear reactors in supplying energy.
    c. There is an abundance of nuclear fuel.
    d. The Tokomak is a nuclear fusion device that is under development.

Mind Expanders

1. Show how each of the following foods is dependent on sunlight.

    a. fish　　　　　b. steak　　　　　c. milk
    d. bread　　　　e. chicken　　　　f. fruits

2. Why do engineers say that inexpensive and easily obtained energy forms are most difficult and expensive to use?

3. How are engineers changing matter into energy?

4. Suggest a solution to the energy problem and give reasons to support your suggestion.

5. Explain how the stars and the sun generate energy.

6. List five energy alternatives to the fossil fuels.

7. Discuss strengths and weaknesses associated with each energy source.

8. Name three types of nuclear radiation and give their:

   a. Greek symbol   b. charge   c. mass

9. Explain two ways that radiation can damage cells.

10. Discuss the advantages and disadvantages of nuclear power plants.

Completion Questions

Write the word(s) from the list below that will correctly complete the statements that follow. Some words may be used more than once.

oil	nuclear fusion
energy	sun
fossil fuel	food
nuclear fission	coal

1. _____ is the ability to do work.

2. _____ , _____ , and _____ are forms of stored sunlight.

3. The source of almost all energy on the earth is _____ .

4. The most expensive fuel is _____ .

5. The easiest fuel to use today is _____ .

6. Coal, gas, and oil are all _____ .

7. The source of all energy on the sun is _____ .

8. _____ is the form of energy that we may depend upon in the immediate future.

9. _____ is the fossil fuel that causes the most pollution.

Find the Facts

Locate the section in this chapter that supports each answer to the completion questions.

True or False

If the statement is true, mark it *T*. If the statement is false, change the indicated word(s), using the list below, to make the sentence true.

coal	green plants
carbon fuels	nuclear fusion
energy	oil
fossil fuels	the sun

1. *Matter* is the ability to force an object to move a distance.

2. The energy crisis is caused by the manipulation of the world supply of *uranium.*

3. Almost all energy on this planet is derived from *fossil fuels.*

4. The source of energy for the stars is from *nuclear fusion.*

5. The energy obtained from *food* is more costly than the energy obtained from any other source.

6. The source of all food on the earth is from *wild animals.*

7. All *energy* forms present some problems.

8. The Tokomak is a *nuclear fusion* device.

9. Acid rains are the result of the use of *geothermal energy* sources.

10. The extensive use of *nuclear energy* may produce excessive carbon dioxide in the air.

Matching

Write the number in column B that relates to the word(s) in column A in the space indicated.

COLUMN A	COLUMN B
_____ solar energy	1. An energy source that is derived from the remains of ancient life forms.
_____ geothermal energy	2. Energy that is derived from the sun.
_____ atomic fission	3. A material from which energy can be derived.
_____ fuel	4. The heat derived from the earth's interior.
_____ nuclear reactor	5. A device that releases controlled energy from the fissioning of atoms.
_____ energy	6. The splitting of atoms for the release of energy.
_____ fossil fuel	7. Putting atomic nuclei together with a release of energy.
_____ atomic fusion	8. The ability to force motion.

Multiple-Choice

1. The most expensive energy source is:
   - a. coal
   - b. electricity
   - c. sugar
   - d. sirloin steak

2. The first source of energy in our universe is:
   - a. the green plant
   - b. solar energy
   - c. nuclear energy
   - d. fossil fuels

3. The most polluting fuel, causing the greatest number of premature deaths, is:
   - a. coal
   - b. oil
   - c. nuclear energy
   - d. wood

4. The time needed to bring a nuclear reactor into operation is:
   - a. 2 years
   - b. 10 years
   - c. 5 years
   - d. 15 years

5. The main problem with solar energy is:
    a. it produces acid rains.
    b. it is difficult to dispose of its unused fuel.
    c. it is too diffuse.
    d. it takes too long to establish the installation.

6. Which element captures a neutron to undergo fission?
    a. Cr-52   b. Sr-88   c. U-235   d. Fe-56

7. A radioisotope used to pinpoint a brain tumor is:
    a. C-12   b. Pb-206   c. Tc-99   d. U-238

8. Which of the following has a negative charge?
    a. alpha particle   b. gamma ray
    c. aluminum ion   d. beta particle

9. Which particle is electrically neutral?
    a. proton   b. electron
    c. neutron   d. alpha particle

Half-Life Problems

(The half-lives, listed in Table 7-8, may be used for the following problems.)

Sample Problem

2 milligrams of cobalt-60 are accidentally injected into a cancer patient. If the cobalt is not excreted, how much of it will remain in the patient after 10.6 years?

> #### Solution
>
> According to Table 7-8, the half-life of Co-60 is 5.3 years, therefore 10.6 years represent two half-lives. After one half-life (5.3 years), half of the Co-60 will have decayed, and 1 mg will remain. At the end of the second half-life (10.6 years) the patient will retain 0.5 mg.

Solve the following problems

1. A doctor administers 12 mg of I^{131} in a sodium iodide (NaI) capsule to a patient. How much iodine remains in the patient after 16 days (assuming there are no losses by excretion)?

2. A green plant takes up 40 g of P^{32} in the form of sodium phosphate (Na_3PO_4). How many days are needed before just 5 grams remain?

3. A museum curator, in checking the authenticity of papyrus from an Egyptian pyramid, burned a 1-gram sample. He found the carbon-14 content in the carbon dioxide to have a count rate of 7.5 counts per minute (7.5 cpm). The carbon-14 content of new papyrus is 15 counts per minute (15 cpm). Is the papyrus genuine? How old is it?

4. A patient drinks a potassium chloride solution that contains 6 mg of K^{42}. How many milligrams will be retained by the patient after one day, assuming no excretions?

5. A baby is fed milk in a formula that contains 4 mg of strontium 90. How many milligrams will the person have at age 56?

Transmutation Equations

Find the unidentified particle in the following transmutation reactions. See section 7-8.

1. $_{84}Po^{210} \longrightarrow {}_{82}Pb^{206} \quad + \quad X$

2. $_{94}Pu^{240} \longrightarrow {}_{95}Am^{240} \quad + \quad X$

3. $_{38}Sr^{90} \longrightarrow X \quad + \quad {}_{-1}e^{0}$

4. $_{88}Ra^{226} \longrightarrow X \quad + \quad {}_{2}He^{4}$

5. $_{45}Rh^{107} \longrightarrow X \quad + \quad {}_{-1}e^{0}$

Careers in Energy

High School Diploma	2-Year Degree	4-Year or Graduate Degree
	Coal Industry	
railroad worker	lab technician	chemist
truck driver	engineering technician	mining engineer
laborer		mine safety inspector
miner		environmental engineer
	Oil Industry	
oil well driller	drilling superintendent	chemist
oil pumper		chemical engineer
welder	lab technician	petroleum engineer
iron worker	engineering technician	geologist
	Nuclear Industry	
plumber	reactor operator	nuclear engineer
electrician	laboratory technician	electrical engineer
pipefitter		nuclear physicist
welder		radiation chemist
carpenter		health physicist
	Solar Energy	
solar panel installer	lab technician	civil engineer
plumber	engineering technician	architect
metal worker		environmental engineer
electrician	environmental technologist	
welder		

Chapter 8

What Is the Chemistry of Construction Materials?

Instructional Objectives

After completing this chapter, you will be able to:

1. Compare construction materials of ancient civilizations with those used in society today.
2. Define cement, ceramics, clay, concrete, diatomaceous earth, glass, limestone, marble, mineral, mortar, plaster of paris, rock, sandstone, silica, and slag.
3. Compare the chemical content of the earth, the human body, and the universe.
4. Explain how cement is made and used.
5. Compare cement, concrete, and reinforced concrete.
6. Explain present uses and predict future uses for ceramics.
7. Explain how glass is made.
8. Explain how mortar holds bricks together.
9. Distinguish between plaster of paris and gypsum.

Chapter 8 Contents

Introduction   **179**

8-1 What Are Construction Materials Made Of?   **182**

8-2 How Is Cement Made?   **184**

8-3 How Are Ceramics Made?   **186**

8-4 How Is Glass Made?   **188**

8-5 How Does Mortar Hold Bricks Together?   **189**

8-6 How Is Plaster Made?   **190**

Written Exercises   **192**

Career Information   **197**

Chapter 8

What Is the Chemistry of Construction Materials?

Introduction

On April 12, 1981, the first reusable space shuttle, **Columbia**, was launched. There was concern on this first test flight that some ceramic (clay) tiles had broken off the belly of the ship, and that the great heat of reentry would melt the aluminum undersurface. If this had happened, the entire ship, including its two-man crew, would have been incinerated (burned up). The ceramic tile on this most advanced technology was basically the same as the clay bricks used by the ancient Egyptians. Many buildings constructed more than 5,000 years ago from clay, limestone, cement, glass, and granite are still standing today. The materials used in construction today are essentially the same as those used thousands of years ago. Limestone, cut from the earth, has value in the construction of buildings. Limestone quarries are found in Vermont and New Hampshire. Italian marble is valued because of its beauty when polished. Marble is an older form of limestone.

Courtesy: Egyptian Government Tourist Authority

FIGURE 8-1a

Construction materials used by the Egyptians have changed very little.

FIGURE 8-1b

Courtesy: Italian Government Travel Office

Roman structures used brick very similar to those in use today.

FIGURE 8-2a

The heat shield on the space shuttle is made of ceramic (clay) tiles.

Courtesy: Goodyear Tire and Rubber Co.

FIGURE 8-2b

Caves are formed in limestone areas. Carbon dioxide, dissolved in rainwater, forms carbonic acid.

$$H_2O + CO_2 \longrightarrow H_2CO_3$$

This dilute acid solution dissolves the limestone into soluble calcium bicarbonate.

$$H_2CO_3 + CaCo_2 \longrightarrow Ca(HCO_3)_2$$

In this photo, a **stalactite** (limestone hanging from the ceiling of the cave) can be seen. They are formed when calcium bicarbonate gives up its water to form calcium carbonate again.

$$Ca(HCO_3)_2 \longrightarrow CaCO_3 + H_2O$$

Stalagmites also form on the floor of the cave. These structures reach upward. When they meet the stalactite, a column is formed.

Table 8-1	
Construction Materials Containing Silicon	**Construction Materials Without Silicon**
cement	limestone
clay	marble (older limestone)
fieldstone	metals
mica or Manhattan schist	plaster of paris
sandstone	
slate	
mortar	

8-1   What Are Construction Materials Made Of?

Our buildings, bridges, dams, and roads are made from materials we find in the crust of the earth. A rock is a mixture of minerals. A **mineral** is a natural chemical compound. Salt and sand are familiar minerals. More than 80% of the earth's solid surface is made of compounds of oxygen, silicon, and aluminum. These elements are the mainstay of our construction materials. Silicon is the second most abundant element on the earth. Aluminum is the most abundant metal on earth. Sand is silicon dioxide, SiO_2, or **silica**. The cell walls of microscopic water plants, called **diatoms**, are made of silica. The remains of these plants, in California, Nevada, and Florida, form diatomaceous earth or **diatomite**. It is used for filters and as an abrasive. Almost all stone, except limestone, contains quartz crystals. Quartz is silicon dioxide. Opal and agate are made of silicon dioxide with trapped impurities to give it special qualities. Opal gets its "fire" from trapped water in its crystalline structure which serves to increase its value.

Table 8-2

Element	Percentage
Percent Composition of the Earth	
oxygen	46.43
silicon	27.77
aluminum	8.14
iron	5.12
calcium	3.63
sodium	2.85
potassium	2.60
magnesium	2.09
hydrogen	0.63
titanium	0.13
chlorine	0.06
carbon	0.03
all others	0.52
Percent Composition of the Universe	
hydrogen	92.8
helium	7.1
others	0.1
Percent Composition of the Human Body	
hydrogen	63.2
oxygen	25.6
carbon	9.5
nitrogen	1.3
phosphorous	0.2
others	0.2

8-2 How Is Cement Made?

Cement is a mixture of clay and limestone. The Egyptians used a cement that looked like plaster of paris when they built the pyramids.

Table 8-3

---

clay + limestone ⟶ cement + calcium silicate

$2\ AlHSiO_4 + 5\ CaCO_3 \longrightarrow Ca_3(AlO_3)_2 + 2\ CaSiO_3 + 5\ CO_2 + H_2O$

In the United States, we use Portland cement. This cement was first made in Leeds, England. It was called Portland cement because when it sets (hardens), it looks like the limestone that comes from Portland Isle in England. It is made by burning a complex mixture of clay, limestone, and similar materials. Cement is made by first grinding the clay into a fine powder, then mixing it with finely ground limestone. The mixture is then put into a rotary kiln (turning oven). This is a long, steel rotating oven lined with fire brick. It is set higher on one end and has **clinkers** on its inside. The clinkers pick up and drop the mixture as the steel cylindrical oven turns. The clay-limestone mixture is carried slowly to the lower end of the kiln by gravity. Hot flames shoot into the kiln from burning gas or coal. The chemical reactions result in Portland cement, as shown above. Cement is appropriate for structures that are constructed in water because it will set under water. Concrete is cement that has small stones or pebbles mixed into it. If steel rods are embedded in the mixture, the strength will be multiplied. This is reinforced concrete. In Hiroshima and in Nagasaki, buildings of reinforced concrete withstood the blast of the atomic bombs in World War II. Nuclear reactors are constructed of reinforced concrete.

Table 8-4

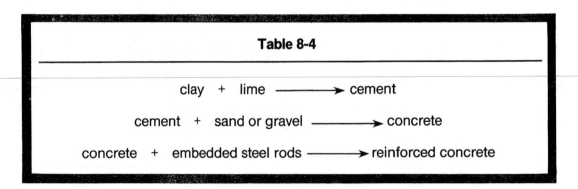

clay + lime ⟶ cement

cement + sand or gravel ⟶ concrete

concrete + embedded steel rods ⟶ reinforced concrete

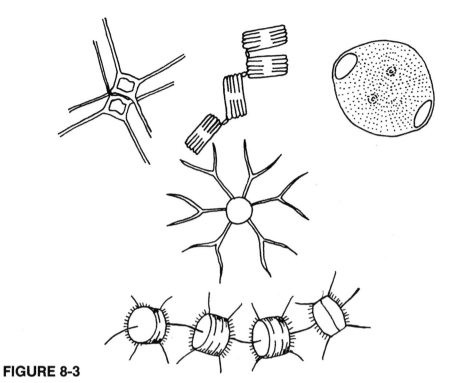

FIGURE 8-3

Diatoms form the slimy coatings on underwater rocks, boats, and piles.

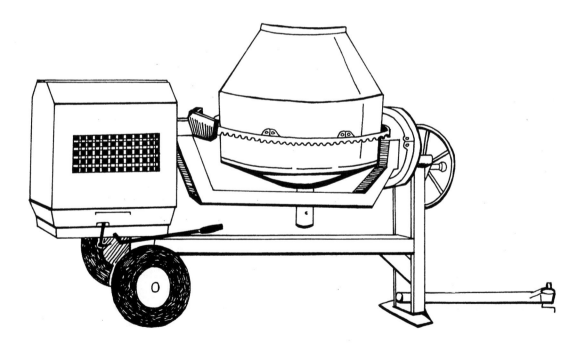

FIGURE 8-4

Cement is made from a fired mixture of limestone and clay.

High-quality cement has **slag** ($CaSiO_3$) (which is obtained from the blast furnace) added to it (see section 9-4). Cement will set in a few hours, but it may take a year before it will achieve its maximum strength. Each component in cement hardens in its own time.

8-3   How Are Ceramics Made?

Almost every civilization has found it necessary to master techniques of molding, shaping, and constructing objects in clay. **Ceramics** is the art of creating useful and artistic clay objects. Pure clay, or **kaolin**, $(H_2O)_2 . Al_2O_3 . 2SiO_2$, is used for the making of fine china and porcelain. The bricks used in the construction of homes and buildings are ceramic. Clay is formed by the decomposition of rocks that contain the mineral called **feldspar**: potassium aluminum silicate ($KAlSi_3O_8$).

Table 8-5

$$2\ KAlSi_3O_8\ +\ 2\ H_2O\ +\ CO_2 \longrightarrow K_2CO_3\ +\ H_4Al_2Si_2O_9\ +\ 4\ SiO_2$$

feldspar + water + carbon dioxide $\longrightarrow$ potash + kaolin + silica

---

When clay is wet, it is plastic and can easily be shaped. It is then fired in an oven or a kiln at high temperature. The American Indians were able to fire their ceramics by burying their clay objects in a pile of animal manure. The manure would be set afire. The pile of manure burned slowly and at a sufficiently high temperature to fire the clay. After the first firing, the earthenware piece became smooth but porous. It would not soften again. The clay object may take several glazes, but each glaze must be fired every time. Glazes usually consist of metallic oxides that vaporize in the kiln, and fill the pores. Cobalt oxide fires to blue, iron to yellow or gold, and chromium to green.

Recent advances made by chemists in ceramic research have produced materials that are stronger than steel, light in weight, and capable of withstanding the heat of a reentering space vehicle or the heat of a blast furnace. High-tech ceramics can be found in fishhooks, batteries, bearings, electronic components, solar cells, artificial bones, artificial teeth, and cutting tools. Their diamond-like hardness makes them resistant to wear and heat. They cannot rust. Research is under way to produce an inexpensive, small, light, and powerful ceramic automobile engine. It is expected to run hotter than cast iron engines. The high temperature will ensure better fuel combustion and improve efficiency by about 30%.

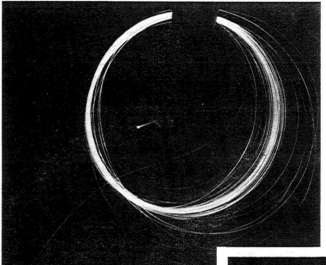

FIGURE 8-5a

Loops of a hair-thin glass fiber, illuminated by laser light, represent the transmission medium for light-wave systems. Typically, 12 fibers are embedded between two strips of plastic in a flat ribbon, and as many as 12 ribbons are stacked in a cable that can carry more than 40,000 voice channels.

Courtesy: AT&T Bell Laboratories

FIGURE 8-5b

Fibers are drawn from glass preforms such as the one held here. Red laser light projected through the side of a preform (wavy lines) is used to determine its optical quality.

Courtesy: AT&T Bell Laboratories

Courtesy: AT&T Bell Laboratories

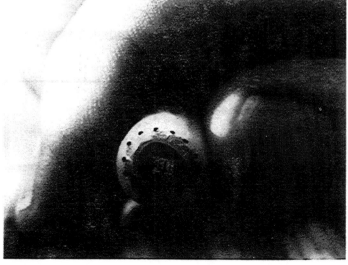

FIGURE 8-5c

A fiberguide cable, similar to the one shown in the photo, carried voice, television, and data signals in the Atlanta experiment. The cable, protected with steel wires embedded in its sheath, contains 12 ribbons, each encapsulating 12 glass fibers, or lightguides. In the Chicago evaluation, 24 lightguides in a half-inch diameter cable will be threaded through existing telephone ducts beneath city streets.

8-4   How Is Glass Made?

Glassmaking is also an ancient art well known to the ancient Egyptians. Glass jewelry can be seen in most museums that exhibit ancient art. Glassmaking saw some progress in the Middle Ages. Venetian glass is valued and hard to match today. The Bell Telephone Company has developed supertransparent optical fibers. They send thousands of telephone conversations along a beam of light on a glass thread. Wire is saved and efficiency is increased. Nuclear scientists find that nuclear wastes can be safely disposed of if they are embedded in glass marbles, then buried deep in salt mines (section 7-15). Glass is made from a melted mixture of sand, sodium carbonate, and calcium carbonate (limestone). When the melted mass cools, we have glass.

Table 8-6

$$6\,SiO_2 \;+\; Na_2CO_3 \;+\; CaCO_3 \longrightarrow Na_2O.CaO.6SiO_2 \;+\; 2\,CO_2$$

sand + sodium carbonate + limestone ⟶ glass + carbon dioxide

Before it hardens, glass may be blown or shaped into any desired object. Glass wool, spun glass, or fiberglass is made by blowing the molten glass through small holes with a steam jet. Impurities added to molten glass will give color to the glass. Iron makes glass green. Blue glass contains cobalt. Selenium makes red glass and is used to make red plastic lenses for automobile taillights. Construction glass is used in the building industry. Special glasses have been developed to resist heat, and are suitable for use in baking foods.

If sand is melted with sodium and potassium oxides, water glass will result. This glass will dissolve in water. It is sometimes added to cement.

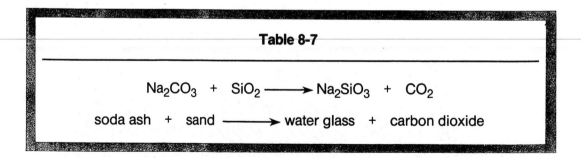

Table 8-7

$$Na_2CO_3 \;+\; SiO_2 \longrightarrow Na_2SiO_3 \;+\; CO_2$$

soda ash + sand ⟶ water glass + carbon dioxide

8-5 How Does Mortar Hold Bricks Together?

The first recorded civilization, the Sumerian, constructed large buildings with clay (adobe) bricks that were set in bitumen (tar or asphalt). Their buildings were eroded away by floods and crumbled with time. The ancient Egyptians and Greeks constructed large structures by placing stones on each other. Their constructions remain to this day. The Romans were the first people to use mortar to hold stones and bricks together. Archaeologists can always identify Roman construction by the mortar between the stones.

A simple mortar may consist of **slaked lime** [$Ca(OH)_2$] and sand (SiO_2). In practice, cement is used in place of calcium hydroxide. Mortar, unlike cement, can set only in the air. As time passes, the mortar develops a stronger hold on the bricks. When the mortar is put on the porous clay brick, the calcium hydroxide solution fills the pores and the spaces between the sand particles in the mortar. In time, the water evaporates. The carbon dioxide in the air changes the calcium hydroxide into calcium carbonate (limestone). The rough surface of the brick is now firmly held to the mortar.

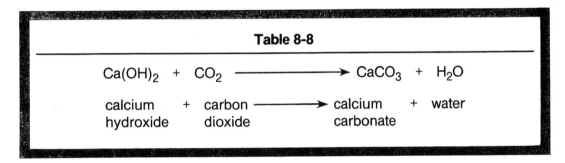

Table 8-8

$$Ca(OH)_2 + CO_2 \longrightarrow CaCO_3 + H_2O$$

calcium hydroxide + carbon dioxide ⟶ calcium carbonate + water

FIGURE 8-6

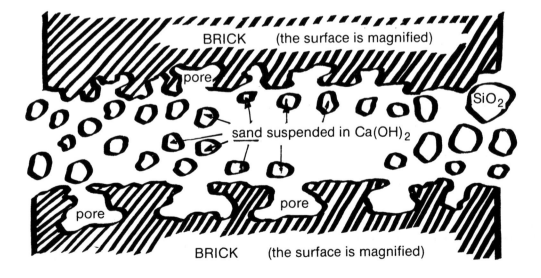

The surface of the brick is rough and has many pores. The pores fill with calcium hydroxide and slowly change into calcium carbonate (limestone). When mortar sets, the bricks are held firmly in place.

8-6   How Is Plaster Made?

Plaster of paris is a fine white powder that will set when mixed with water. It can be poured into molds and made into different shapes. After the doctor sets a broken bone, he may wrap the limb in bandages that have been dipped into a plaster of paris paste. When the plaster sets, a rigid cast is formed. Dentists may use plaster of paris to record an impression of your teeth. The walls of your classroom, and those of your home, are probably made of plaster. Plaster walls insulate your living quarters from noise. Plaster also insulates homes from heat in the summer and cold in the winter. Plaster of paris is calcium sulfate: $(CaSO_4)_2.H_2O$. When mixed with water, the plaster will form a paste.

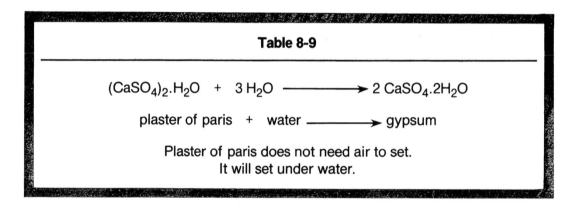

Table 8-9

$$(CaSO_4)_2.H_2O \quad + \quad 3 H_2O \longrightarrow 2 CaSO_4.2H_2O$$

plaster of paris   +   water ⟶ gypsum

Plaster of paris does not need air to set.
It will set under water.

If you started with plaster walls at home, you are now living inside gypsum walls. Plaster of paris sets to become this hard mineral. When gypsum is subjected to intense heat, it changes back to plaster of paris, as its water is driven off.

Table 8-10

$$2 CaSO_4.2H_2O \quad — \quad HEAT \longrightarrow (CaSO_4)_2.H_2O \quad + \quad 3 H_2O$$

gypsum   —   HEAT ⟶ plaster of paris   +   water

Now You Know

1. Construction materials have changed very little from those used in ancient civilizations.

2. Oxygen is the most abundant element on the surface of the earth, followed by silicon, and then aluminum. These three elements account for more than 80% of the earth's solid surface.

3. Oxygen, silicon, and aluminum are found in all cements, clays, stones, sand, and mortar.

4. Cement is a mixture of calcium silicate and calcium aluminum silicates. It is made from clay and limestone.

5. Concrete is cement with the addition of sand or small stones.

6. Reinforced concrete is concrete in which steel rods are embedded.

7. Rocks are mixtures of minerals.

8. Minerals are natural chemical compounds.

9. Clay is made from the mineral feldspar.

10. Glass is the melted mixture of sand (SiO_2) and metallic oxides or carbonates. When sodium and potassium oxides are melted with sand, water glass is produced.

11. Mortar is a mixture of sand with calcium hydroxide or cement. When mortar sets, it loses its water to evaporation. The calcium hydroxide reacts with carbon dioxide to form hard, stonelike calcium carbonate.

12. Plaster of paris is made by driving off water from gypsum under intense heat. Plaster of paris will set under water. It does not need air.

New Words

cement	A complex mixture of limestone and clay.
ceramics	Another name for clay.
clay	Aluminum silicate. It is made from feldspar. It is plastic when wet and hard and brittle when fired in a kiln.
concrete	A mixture of cement with sand or small stones.
diatomaceous earth	The earthy remains of diatoms. It is used for filters and abrasives.

glass	A mixture of sodium and calcium silicates.
limestone	A rock composed of calcium carbonate.
marble	An older limestone.
mineral	A natural chemical compound.
mortar	A mixture of sand and cement or slaked lime. It is used to hold bricks together.
plaster of paris	A calcium sulfate compound that reacts with water to set into the hard mineral gypsum.
rock	A mixture of minerals
sandstone	A rock composed of small grains of sand naturally cemented together.
silica	Sand or silicon dioxide: SiO_2
slag	Calcium silicate. It is sometimes added to cement. It is a product of the blast furnace in the smelting of iron.

Reading Power

For each of the following questions, select one answer that seems most correct.

1. The main idea of the introduction is:
   a. Ceramic tiles are used as a heat shield for spaceships.
   b. The ancient Egyptians and Romans built structures that are still standing today.
   c. The materials used for construction today have not changed much from those used thousands of years ago.
   d. There have been great changes in the nature of construction materials over the years.

2. The main idea of section 8-1 is:
   a. Most of our common building materials are compounds of oxygen, silicon, and aluminum.
   b. Oxygen is the most abundant element on the surface of the earth.
   c. Semiprecious stones such as opal and agate are composed of silicon compounds.
   d. Diatoms are one-celled marine plants that are composed of silicon dioxide.

3. The main idea of section 8-2 is:

    a. Cement is made from a mixture of limestone and clay.
    b. Cement has been in use for over 5,000 years.
    c. Steel rods embedded in concrete multiply the strength of the concrete structure.
    d. Concrete is a mixture of cement and sand.

4. The main idea of section 8-3 is:

    a. Clay and ceramics mean the same thing.
    b. Ceramics are made from the mineral feldspar.
    c. Ceramics are developing new uses that make them more important in the space age.
    d. Ceramic engineering is a rewarding career field.

5. The main idea of section 8-4 is:

    a. Glass had been made by the ancient Egyptians.
    b. The Venetians in the Middle Ages made great advances in glassmaking.
    c. The telephone company is using glass fibers instead of wires to carry telephone conversations.
    d. Glass is made by melting sodium and calcium oxides with sand.

6. The main idea of section 8-5 is:

    a. Mortar is a mixture of sand and calcium hydroxide.
    b. When mortar sets, its water evaporates and calcium hydroxide reacts with carbon dioxide to form calcium carbonate.
    c. Mortar holds bricks together with its rough and porous surface.
    d. Mortar could not hold smooth, glazed bricks together.

7. The main idea of section 8-6 is:

    a. Plaster of paris will set under water.
    b. Plaster of paris can be made from the mineral gypsum.
    c. Plaster of paris is used by doctors to make casts for the healing of broken bones.
    d. The walls of your classroom started as plaster of paris, but have set into gypsum.

Mind Expanders

1. Describe, in three or four sentences, how concrete is made.

2. What is glaze? What purpose does it serve?

3. How have construction materials remained the same today as they were 5,000 years ago?

4. How have the uses of glass and ceramics changed in recent times?

5. Explain how mortar sets in air.

6. Explain how Portland cement is made.

7. What is the difference between a rock and a mineral? Name two examples of each.

Completion Questions

Write the word(s) from the list below that will correctly complete the following statements.

aluminum	mortar
cement	oxygen
concrete	plaster of paris
glaze	silicon

1. _____ is a fine powder formed by grinding and heating clay and limestone.

2. _____ , _____ , and _____ are the three most abundant elements in the crust of the earth.

3. _____ is applied to a fired ceramic to give it a gloss and to seal its pores.

4. Gypsum becomes _____ after its water is driven off.

5. _____ is a mixture of sand and cement.

Find the Facts

Choose the section in this chapter that supports each answer to each completion question.

True or False

If the statement is true, mark it *T.* If the statement is false, change the indicated word(s), using the list below, to make the statement true.

aluminum	feldspar	glass
mortar	steel rods	

1. *Mortar* is made from a mixture of slaked lime and sand.

2. Reinforced concrete is concrete to which *pebbles* have been added.

3. The mineral *quartz* is used to make clay.

4. The most abundant element on earth is *oxygen*.

5. *Clay* is the molten mass of sand and sodium oxide.

Matching

Write the number in column B that relates to the word(s) in column A, in the spaces indicated.

COLUMN A		COLUMN B
_____ mineral	1.	A construction material made from clay and calcium carbonate.
_____ silica	2.	A mixture of sand and calcium hydroxide.
_____ rock	3.	Prepared by heating gypsum.
_____ clay	4.	Made of clay and limestone.
_____ cement	5.	A natural chemical compound.
_____ Portland cement	6.	Prepared from feldspar. It is plastic when wet, and brittle when fired.
_____ mortar	7.	Sand SiO_2.
_____ plaster of paris	8.	A natural form of calcium carbonate.
_____ limestone	9.	Obtained from a blast furnace and used in some cements.
_____ slag	10.	A mixture of minerals.

Multiple-Choice

1. Cement is a complex mixture of:
    a. clay and limestone
    b. limestone and plaster of paris
    c. sand and slaked lime
    d. clay and sand

2. Construction materials:

    a. are basically different from those of the past.
    b. are basically the same as those of the past.
    c. lack the element silicon.
    d. lack the element oxygen.

3. The elements, in their order of abundance, are:

    a. oxygen, silicon, iron.
    b. silicon, oxygen, aluminum.
    c. oxygen, silicon, aluminum.
    d. aluminum, silicon, oxygen.

4. Ceramics are finding new uses today as:

    a. bricks for the construction industry.
    b. tiles for rooftops.
    c. making pieces of art.
    d. heat shields in the space industry.

5. Glass is finding new uses today in:

    a. mirrors   b. jewelry
    c. beads   d. optical fibers

Careers in the Construction Industry

High School Diploma

asbestos/insulating worker, bricklayer, carpenter, cement mason, electrician, glazier, laborer, marble setter, painter, paperhanger, plasterer, plumber, roofer, stonemason, terrazzo worker, tile setter, insurance inspector, welder, etc.

2-Year Degree

surveyor
operating engineer
construction machine operator
fire inspector
construction inspector

4-Year or Graduate Degree

civil engineer
electrical engineer
mechanical engineer
insurance underwriter

Chapter 9

How Are Metals Obtained?

Instructional Objectives

After completing this chapter, you will be able to:

1. Define activity (of metals), alumina, blister copper, bronze, calcine, case hardening, cast iron, charge, copper matte, cryolite, flotation, flux, gangue, ore, pig iron, quenching, reduction, reverberatory (furnace), roasting, slag, steel, and tempering.

2. Explain the difference between first-rate and second-rate ores.

3. List one ore for each of the following metals: iron, copper, and aluminum.

4. Explain how iron, lead, copper, aluminum, and titanium are extracted from their ores.

5. Compare three types of steel.

6. Outline the steps of a generalized metallurgical process, and specific steps for iron, copper, aluminum, and titanium.

7. Compare the handling of oxide, sulfide, and carbonate ores.

8. Compare steel with pig iron.

9. Recognize that titanium is a modern material developed for the jet/space age.

Chapter 9 Contents

Part I

	Introduction	201
9-1	Where Are Metals Found?	202
9-2	Why Can't All Metals Be Found Free in Nature?	202
9-3	How Are Metals Extracted from Their Ores?	206
9-4	How Is Iron Extracted from the Earth?	206
9-5	Why Is Steel Preferable to Iron?	209
9-6a	How Is Steel Made?	211
9-6b	How Is Bessemer Steel Made?	211
9-6c	How Is Open-Hearth Steel Made?	211
9-6d	How Is Electric-Furnace Steel Made?	212
9-7	How Is High-Carbon Steel Heat-Treated?	213
9-8	How Is Low-Carbon Steel Heat-Treated?	214
9-9	How Can the Rusting of Steel Be Prevented?	214
	Written Exercises	217

Part II

9-10	How Is Lead Extracted from the Earth?	221
9-11	How Is Copper Extracted from the Earth?	222
9-12	How Is Copper Electrolytically Refined?	223
9-13	How Did Aluminum, Our Most Abundant Metal, Become Available?	225
9-14	How Is Aluminum Liberated from the Earth?	226
9-15	What Is the Metal of Opportunity Today?	227
9-16	How Is Titanium Extracted from Its Ores?	227
	Written Exercises	229
	Career Information	232

Chapter 9

How Are Metals Obtained?

Part I

FIGURE 9-1
Courtesy: U.S. Dept. of Defense

Titanium alloys were developed by metallurgical engineers for use in jet engines. They can withstand the intense heat and extreme forces that these engines produce. This is the rear view of a U.S. Air Force F-4 Phantom jet taking off with its afterburners glowing.

Introduction

In 1978, doctors substituted steel rods for shattered bones and hips. After seven years, the metal developed cracks and failed. Since then, alloys have been developed that outlast human bones. Buses, automobiles, and airplanes have also experienced metal failure. The titanium alloy linings of a rocket or jet engine were developed after many years of intensive research. They must withstand severe forces and high temperatures. Why do some metals succeed where others fail? How do we get metals? What is the job of the metallurgical engineer?

FIGURE 9-2   Courtesy: New York City Transit Authority

The N.Y.C. Transit Authority removed new buses from the streets after five
months of service because metal parts broke due to metal failure.

9-1   Where Are Metals Found?

About 80% of the elements are metals. They are found as parts of natural chemical compounds, or **minerals** (see section 8–1). The rocks of the earth are mixtures of
minerals.

The first metals used by humans were probably gold, silver, and copper. These
metals can be found free in nature. Iron, lead, and other metals required energy and
technology to be liberated from their compounds. The Iron Age came about many
years after the Copper Age (also called the Bronze Age). The ancient techniques for
the production of iron, copper, tin, and lead have changed very little. Aluminum and
titanium were not discovered until modern times. These metals require special technologies that were not available to early civilizations.

9-2   Why Can't All Metals Be Found Free in Nature?

In Chapter 4, we saw that atoms of metals give away their valence electrons.
Some metals lose their electrons more easily than others. The ease with which a
metal releases its valence electrons is called its **activity**.

Table 9-1

Ore/Mineral	Formula	
Iron Ores		
magnetite	Fe_3O_4	
hematite	Fe_2O_3	
limonite	mixture FeO + Fe_2O_3	
taconite	contains hematite and magnetite	
Copper Ores		
native copper	Cu	
chalcocite	Cu_2S	
malachite	$CuCO_3$	
azurite	$CuCO_3$	
cuprite	Cu_2O	
Other Ores		
uraninite	UO_2	
carnotite	mixture of uranium and vanadium	
aluminum ore	bauxite	Al_2O_3
lead ore	galena	PbS
titanium ore	ilmenite	$FeTiO_3$
zinc ore	sphalerite	Zns

Table 9-2

Activity of Metals

(In water at 25°C and 1 atm)

The metals are listed in their decreasing order of activity. The most active metal is listed first and the least active metal is last on the list.

Metal	Ionization Equation	* Volts
lithium	$Li - e \longrightarrow Li^+$	+ 3.04
rubidium	$Rb - e \longrightarrow Rb^+$	+ 2.98
potassium	$K - e \longrightarrow K^+$	+ 2.93
cesium	$Cs - e \longrightarrow Cs^+$	+ 2.92
barium	$Ba - 2e \longrightarrow Ba^{2+}$	+ 2.91
strontium	$Sr - 2e \longrightarrow Sr^{2+}$	+ 2.89
calcium	$Ca - 2e \longrightarrow Ca^{2+}$	+ 2.87
sodium	$Na - e \longrightarrow Na^+$	+ 2.71
magnesium	$Mg - 2e \longrightarrow Mg^{2+}$	+ 2.37
aluminum	$Al - 3e \longrightarrow Al^{3+}$	+ 1.60
manganese	$Mn - 2e \longrightarrow Mn^{2+}$	+ 1.18
zinc	$Zn - 2e \longrightarrow Zn^{2+}$	+ 0.76
iron	$Fe - 2e \longrightarrow Fe^{2+}$	+ 0.44
cobalt	$Co - 2e \longrightarrow Co^{2+}$	+ 0.28
nickel	$Ni - 2e \longrightarrow Ni^{2+}$	+ 0.25
tin	$Sn - 2e \longrightarrow Sn^{2+}$	+ 0.14
lead	$Pb - 2e \longrightarrow Pb^{2+}$	+ 0.13
hydrogen	$H - e \longrightarrow H^+$	0.00
copper(ic)	$Cu - 2e \longrightarrow Cu^{2+}$	– 0.34
copper(ous)	$Cu - e \longrightarrow Cu^+$	– 0.52
mercury(ic)	$Hg - 2e \longrightarrow Hg^+$	– 0.78
mercury(ous)	$Hg - e \longrightarrow Hg^+$	– 0.79
silver	$Ag - e \longrightarrow Ag^+$	– 0.80
gold	$Au - 3e \longrightarrow Au^{+3}$	– 1.50

*Volts can be thought of as a measure of metallic activity.

More active metals replace less active metals in compounds. The atoms of zinc lose their electrons more easily than atoms of copper. When zinc atoms contact copper ions (Cu^{+2}), the zinc atoms force their electrons onto the copper ions. The result is that the zinc ionizes and the copper ion is restored to the free (uncombined) metal.

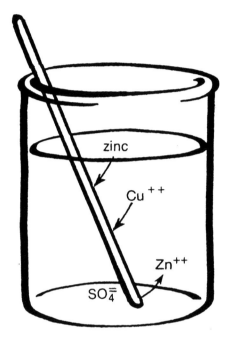

FIGURE 9-3

$$Zn^0 + CuSO_4 \qquad ZnSO_4 + Cu^0$$

More active metals replace less active metals from their compounds. In this beaker, the zinc strip dissolves, and the copper ions deposit on the strip as free metallic copper.

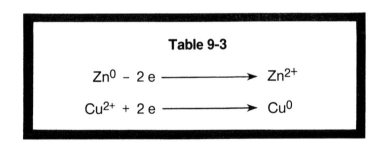

Table 9-3

$$Zn^0 - 2\,e \longrightarrow Zn^{2+}$$

$$Cu^{2+} + 2\,e \longrightarrow Cu^0$$

A free metal is shown with a zero charge. Metallic ions are positively charged due to the loss of valence electrons. Nonmetallic ions are negatively charged due to the gain of electrons.

The activity table of metals shows that gold, silver, and copper are all less active than hydrogen. Since they cannot replace the hydrogen in water or any acid, they may possibly be found free in nature. The metals listed above hydrogen, in the activity series, are all more active than hydrogen and capable of replacing the hydrogen in water and acids. They are almost never found free in nature.

9-3 How Are Metals Extracted from Their Ores?

Metallurgy deals with the freeing of metals from their **ores**. An ore is a mineral-bearing rock. The mineral is a compound of the desired metal. The nonmineral part of the ore is called the **gangue** or waste.

MINERAL + GANGUE = ORE

In general, to separate a metal from its ore, we must complete the following steps.

---

1. Mine the ore.
2. Separate the mineral from the gangue.
3. Change the mineral into the oxide of the metal.
4. Remove the oxygen from the metallic oxide (reduction).
5. Refine the metal (remove the impurities).

9-4 How Is Iron Extracted from the Earth?

The best ores of iron are magnetite and hematite. These ores are considered best because they are richer in minerals and contain less gangue. Unfortunately, the world supply of these excellent ores is about exhausted. In the United States, iron is extracted from second-rate iron ores, such as **taconite** (named after the Taconic Indians). This ore is mined in the Mesabi Range near Duluth, Minnesota. The ore is first concentrated using the **flotation process**. This process was discovered by a washwoman. She noticed that as she laundered the miners' ore- and grease-stained pants, the mineral particles would stick to the oily and soapy bubbles. The pebbles sank to the bottom of the tub. Mining engineers separate the gangue from the mineral by first pulverizing (grinding into a powder) the ore. The powdered ore is then put into an oil-water bath and vigorously frothed up with air. The mineral sticks to the oily bubbles and the froth is skimmed off the surface. The gangue settles to the bottom and is discarded. The mineral consists of iron oxide: Fe_2O_3. This mineral is then mixed with coal and limestone and made into a **charge**. The charge is the mixture put into the blast furnace, to be converted into an impure form of iron called **pig iron** or **cast iron**.

Courtesy of Umetco Minerals Corp.

Courtesy: Union Carbide Co.

FIGURE 9-4a

Flotation cell, a standard tool in mining for 50 years, here carries copper and molybdenum ore particles in an oil bubble froth, while unwanted rock particles sink to cell bottom. Ore is crushed before flotation can take place.

FIGURE 9-4b

Tungsten ore is separated from its waste rock by flotation. Flotation is a physical-chemical method of removing the ore by bubbling air through the slurry. The chemicals allow the tungsten-ore particles to attach themselves to the air bubbles, which carry them to the top of the slurry to form a mineral-rich froth which is skimmed off. The waste rock sinks to the bottom of the flotation cell and is carried away to the tailings pond.

FIGURE 9-5a

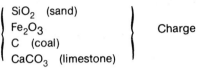

$$SiO_2 \quad \text{(sand)}$$
$$Fe_2O_3$$
$$C \quad \text{(coal)}$$
$$CaCO_3 \quad \text{(limestone)}$$

Charge

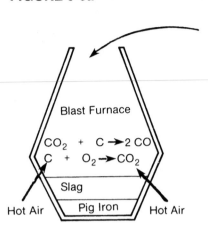

Blast Furnace

$CO_2 + C \rightarrow 2\,CO$
$C + O_2 \rightarrow CO_2$

Slag

Hot Air   Pig Iron   Hot Air

Reactions in the Blast Furnace

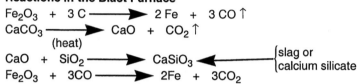

$Fe_2O_3 + 3\,C \longrightarrow 2\,Fe + 3\,CO \uparrow$
$CaCO_3 \longrightarrow CaO + CO_2 \uparrow$
         (heat)
$CaO + SiO_2 \longrightarrow CaSiO_3 \longleftarrow$ {slag or calcium silicate
$Fe_2O_3 + 3\,CO \longrightarrow 2\,Fe + 3\,CO_2$

The slag is used to make cement (see section 8-2). Pig iron is impure.

Courtesy: U.S. Steel Corp.

FIGURE 9-5b

Molten pig iron flowing from a blast furnace.

Courtesy: U.S. Steel Corp.

FIGURE 9-5c

BLAST FURNACE

The carbon in the coal removes the oxygen from the iron oxide. Carbon monoxide also reduces the iron oxide.

The flotation of the ore does not produce a complete separation of the mineral from the gangue. The charge put into the blast furnace still contains much gangue. A **flux** must be added to the charge, in order to eliminate the gangue. If the ore is mined in a sandy area, the gangue will be sand, and the flux will be limestone. When the gangue is limestone, sand is used as the flux to eliminate the gangue as slag.

Table 9-4

$$\text{gangue} + \text{flux} \longrightarrow \text{slag}$$

$$\text{sand} + \text{limestone} \longrightarrow \text{slag}$$

$$SiO_2 + CaCO_3 \longrightarrow CaSiO_3 + CO_2 \uparrow$$

$$2\,CO + O_2 \longrightarrow 2\,CO_2 + \text{heat}$$

Note: Carbon monoxide forms in the blast furnace. It is used as a fuel to heat the incoming air.

Blast furnaces never shut down. They run steadily for 24 hours every day. As the charge is added at the top of the furnace, the molten slag and cast iron are drained off at the bottom.

9-5   Why Is Steel Preferable to Iron?

Cast iron contains many impurities such as carbon, silicon, sulfur, phosphorous, and slag. These impurities make the cast iron very hard, brittle, heavy, and vulnerable to rust. Cast iron cannot be welded. It is not **ductile**. Ductility is a metal's ability to be drawn into a wire. Cast iron is not **malleable**. Malleability is a metal's ability to be rolled or hammered into sheets or plates. This iron can only be cast into molds. It is used to make water mains, automobile engine blocks, and heavy cooking pots and pans. The cast iron from the blast furnace is immediately cast into molds shaped like the trough used by farmers for feeding their pigs—hence the other name, "pig iron." Steel is needed because it does not have the limitations of cast iron. Steel can be welded. It is ductile and malleable. It is lighter in weight than cast iron. It can hold a sharp cutting edge on a knife blade. It is more resistant to rusting. Steel is tougher than cast iron (less brittle).

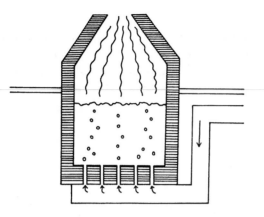

FIGURE 9-6a

Bessemer Converter

Bessemer steel is made by forcing hot air through the holes in the bottom of the furnace for fifteen minutes to burn off impurities.

FIGURE 9-6b

Bessemer Furnaces

Courtesy: U.S. Steel Corp.

Courtesy: U.S. Steel Corp.

FIGURE 9-6c

Oxygen Furnace for "Instant Steel"

9-6a   How Is Steel Made?

Steel is an **alloy** (a solid solution) of iron and carbon, with other metals added to give it special properties. Carbon is soluble in molten iron. In order to make steel, many impurities in the pig iron must be removed. There are three important ways of making steel: the Bessemer process, the open-hearth process, and the electric-furnace process.

9-6b   How Is Bessemer Steel Made?

The Bessemer process, introduced in 1856, produces 15 tons of steel per "burn." The converter has small holes that perforate the bottom of the egg-shaped furnace. Hot air is forced through the molten cast-iron charge for about fifteen minutes to burn off the impurities. The oxides of iron are reduced to iron with carbon and carbon monoxide. A certain amount of carbon and other special elements are added to the iron, to impart special properties to the steel. Usually, a calculated amount of manganese and carbon, in a mixture called **spiegeleisen** is added. "Instant steel" can be made by blasting pure oxygen through the molten charge for fifteen seconds instead of fifteen minutes.

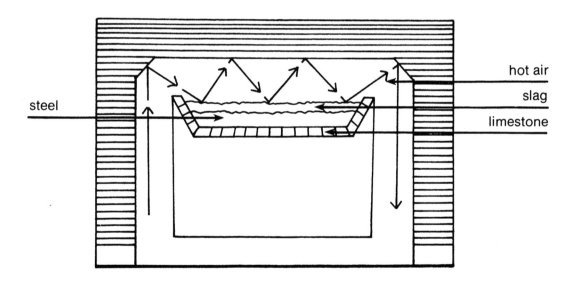

FIGURE 9-7

The open-hearth furnace produces 100 tons of steel in a single heat.

9-6c   How Is Open-Hearth Steel Made?

Most steel is made in an open-hearth furnace. The charge consists of scrap iron, pig iron, and steel. The hearth, or bowl-like bed that holds the charge, is lined with limestone to convert any sand present into slag. There is a swift hot-air flow that changes its direction every ten to fifteen minutes. The purification process takes place for eight to twelve hours. The charge is sampled at all times, so that the quality of the steel is carefully controlled.

FIGURE 9-8

Charging Open-Hearth Furnaces

Courtesy: U.S. Steel Corp.

9-6d   How Is Electric-Furnace Steel Made?

There are different types of electric-furnace steel. Each type is designed to possess special characteristics. Electric-furnace steel is a high-grade steel of a definite alloy. The charge used for the electric furnace is scrap steel instead of pig iron. An electric current is used to melt the charge. The current flows from carbon electrodes in the form of hot arcs or sparks to the charge. The impurities float on the molten surface of the charge. They are skimmed off and alloying metals are added. The formulation of this steel is carefully calculated. One of the special steels made is called **high-speed steel**. This is a vanadium steel used to make power-saw blades. These blades can saw through wood and metal at high speeds. The blade's sharp cutting edge will not warp or dull, even when it gets red hot.

Stainless steel is another valuable product. It will not rust. It will hold its attractive finish because of the chromium that is alloyed into the steel. The disadvantage of electric-furnace steel is that it is expensive.

FIGURE 9-9

Molten High-Grade Steel Flowing from an Electric Furnace

Courtesy: U.S. Steel Corp.

9-7   How Is High-Carbon Steel Heat-Treated?

The old-fashioned blacksmith heats up a piece of steel to soften it, then pounds it into the shape of a horseshoe. The horseshoe is then **quenched**, or cooled rapidly in water. The toughness of the steel is thereby increased. **Tempered steel** is first heated to a high temperature, then cooled rapidly in fresh water, saltwater, oil, or air. Only high-carbon steel, which is usually hard and brittle, can be tempered. Quenching allows the steel to lock in its high temperature structure at the lower (cooler) temperature. The cooling is so rapid that there is no time for the crystalline structure of the high-temperature steel to change back into its low-temperature form. Tempering toughens the high-carbon steel, making it less brittle.

9-8   How Is Low-Carbon Steel Heat-Treated?

Another heat treatment for low-carbon steel is **case hardening**. Low-carbon steel is tough, but soft. Automobile gears sustain much wear on their surfaces. If the low-carbon steel gears are packed with carbon or nitrogen into molds and subjected to high temperatures and high pressures, the carbon or nitrogen will become alloyed to the surface of the metal. As a result, the surface will be hardened, while the inside of the gear will remain tough. The longer the process continues, the thicker the hard skin surface becomes. Gears must be tough so they will resist cracking, but they also need a hard surface, to resist wear.

9-9   How Can the Rusting of Steel Be Prevented?

Rusting takes place when iron or steel combines with oxygen in the air. Rusting is a form of slow oxidation or burning. This reaction takes place spontaneously. It is nature's way of returning the iron to its natural state as it was found in its ores. The reddish-brown coating of iron oxide (**rust**) soon flakes off. When this happens, fresh iron is exposed to oxygen and the iron surface begins to corrode again. The structure is weakened as the iron or steel structure is eaten away. As a result of corrosion, bridges, buildings, and elevated highways collapse. Holes will rust through the steel sides of ships and automobiles. The presence of salt and water catalyzes (speeds) the rusting process. Rusting can be prevented if iron and steel structures are not allowed to come into contact with oxygen.

One way to protect the iron surface from oxygen and salt water is by painting it. The disadvantage is that paint may chip off. If the exposed iron should rust before it is repainted, the rust will spread and lift away the paint until the entire surface is covered with rust.

Machines and tools are often coated with oil when they are stored. The problem with this technique is that the oil is easily removed and the tools will then be subject to corrosion.

Many automobiles are made with galvanized steel. This is steel that is plated with zinc. When oxygen combines with iron, electrons are transferred from the iron to the oxygen (see section 4-2). If the zinc plate is broken, the neighboring zinc atoms compete with the iron to donate their electrons to the oxygen. The zinc corrodes instead of the iron. In this way, zinc (or any metal more active than iron; see Table 9-2) will continue to protect steel against rust even if the protective plate is scratched or broken.

Engineers tried to protect the 800-mile trans-Alaska oil pipeline from corrosion by wrapping the pipe with tape to keep out moisture. In addition, bags of magnesium and zinc were attached to the pipe. Unfortunately, in the rush to finish the pipeline, the workers did not adequately complete its coating. As a result, the pipe is rusting and must undergo costly repairs and replacement.

Now You Know

1. The metallurgical engineer has the responsibility of designing metal alloys that will meet the requirements of the job they must perform.

2. About 80% of the elements in the universe are metals.

3. An ore is a mixture of minerals and gangue.

4. Metals have different activities.

5. Active metals replace less active metals from their compounds.

6. Metals are extracted from their ores in five steps.
    a. Mine the ore.
    b. Use flotation to concentrate the ore.
    c. Change the mineral into an oxide of the metal.
    d. Reduce the oxide.
    e. Refine the metal.

7. Taconite, a second-rate ore of iron, is mined in the United States. It is used because our first-rate iron ore supplies are exhausted.

8. Cast iron is an impure form of iron obtained from the blast furnace.

9. Steel has many advantages over cast iron.

Properties of Cast Iron	Properties of Steel
heavy	lighter
brittle	tougher
not malleable	malleable
not ductile	ductile
rusts easily	more resistant to rust
cannot be welded	can be welded
must be cast into molds	can be shaped

10. Steel is an alloy of iron and carbon, with controlled impurities added for special properties.

11. There are three different ways to make steel.
    a. Bessemer process
    b. open-hearth process
    c. electric-furnace process

12.   The properties of steel can be controlled by heat treatment.

13.   High-carbon steel can be tempered.

14.   Low-carbon steel can be case hardened.

New Words

activity	Metallic activity is a measure of the ease with which metallic atoms release their valence electrons when they react.
case hardening	The alloying of a low-carbon steel with carbon under high pressure and high temperature. The result is a hard surface with a tough interior.
cast iron	The impure iron that is the product of a blast furnace. It is another name for pig iron.
charge	The mixture of materials put into a blast furnace.
combustion	The combination of materials with oxygen. The process of burning.
corrosion	The slow eating away of iron or steel structures due to the spontaneous combination of iron or steel with oxygen. The process of rusting.
flotation	A method by which the mineral is separated from the gangue. The purpose is to concentrate the mineral.
flux	The material added to the charge to eliminate the gangue as slag.
gangue	The waste or nonmineral part of the ore.
ore	A mineral-bearing rock from which a metal is extracted.
pig iron	Another name for cast iron.
quenching	Rapid cooling after heating.
reduction	The removal of oxygen from a metallic oxide.
rust	The reddish-brown coating of iron oxide formed when iron or steel combines with oxygen. See *corrosion*.
slag	Calcium silicate, formed by the reaction of the gangue and the flux.

steel   An alloy of iron and carbon. Other impurites may be added to produce special properties.

tempering   Heating and then quenching of high-carbon steel to obtain special properties.

Reading Power

For each of the following questions, select one answer that seems most correct.

1. The main idea of the introduction is:

  a. Intense heat and enormous force press on the metallic walls of rocket and jet engines.
  b. New buses and jet airplanes experienced mechanical failures because of metallurgical defects.
  c. Metallic alloys were developed to outlast human bones.
  d. We can expect to learn how we obtain metals in Chapter 9.

2. The main idea of section 9-1 is:

  a. Metals are found as parts of minerals in the earth.
  b. Some metals are found free in nature.
  c. After the Stone Age came the Bronze Age.
  d. About 80% of the elements are metals.

3. The main idea of section 9-2 is:

  a. The activities of different metals are not the same.
  b. Atoms of metals lose their valence electrons.
  c. Active metals replace the less active metals in compounds.
  d. Metals less active than hydrogen can be found in nature.

4. The main idea of section 9-3 is:

  a. An ore is composed of a mineral and gangue.
  b. The mineral is the valuable part of the ore because it contains the desired metal.
  c. Active metals replace less active metals in compounds.
  d. In general, the five steps listed in section 9-3 outline the procedure for liberating metals from their ores.

5. The main idea of section 9-4 is:

  a. The five steps in section 9-3 apply to the extraction of iron from its ore.
  b. We are mining second-rate iron ores in the United States because we used up our first-rate ores.
  c. A first-rate ore means that the ore is richer in mineral and poorer in gangue.
  d. The flux changes the gangue into slag.

6. The main idea of section 9-5 is:

   a. Steel can be welded, but cast iron cannot be welded.
   b. Cast iron has limited usefulness because of its impurities.
   c. Steel has properties superior to those of cast iron.
   d. Steel is ductile and malleable.

7. The main idea of section 9-6a is:

   a. Steel is an alloy of iron and dissolved carbon.
   b. Steel is made by controlling the impurities in the alloy.
   c. There are three different ways to make steel.
   d. Carbon is soluble in molten iron.

8. The main idea of section 9-6b is:

   a. The molten charge of cast iron is changed into steel in about fifteen minutes.
   b. Bessemer steel produces about fifteen tons of steel per burn.
   c. Bessemer steel is made in a Bessemer converter.
   d. Spiegeleisen is the alloying impurity added to Bessemer steel.

9. The main idea of section 9-6c is:

   a. Open-hearth steel is made from scrap metal in an open-hearth furnace.
   b. Most steel made today is open-hearth steel.
   c. Open-hearth steel has better quality control because it can be watched for eight to twelve hours.
   d. Open-hearth steel produces about 100 tons per burn.

10. The main idea of section 9-6d is:

    a. Electric-furnace steel is used to make tools.
    b. Electric-furnace steel starts with scrap steel.
    c. The heat, to melt the charge, is from electricity.
    d. All of the above.

11. The main idea of section 9-7 is:

    a. The properties of steel can be changed by the way it is heated and cooled.
    b. Tempered steel is heated to a high temperature.
    c. Tempering steel means that the steel has a soft, tough interior, and a hard surface.
    d. Quenching means rapid cooling.

12. The main idea of section 9-8 is:

    a. Low-carbon steel is treated by case hardening.
    b. Low-carbon steel is tough but soft.
    c. Low-carbon steel can be tempered.
    d. Gears are made of case-hardened steel.

13. The main idea of section 9-9 is:

    a. The Alaska pipeline is rusting because of faulty workmanship.
    b. When iron rusts, its surface is eaten away.
    c. Rusting can be prevented by not allowing iron or steel to contact oxygen.
    d. Zinc protects iron from rust.

Mind Expanders

1. Why did the Bronze Age come before the Iron Age?

2. Why are metals more active than hydrogen never found free in nature?

3. Why can a low-carbon steel never be tempered?

4. How does case hardening make the surface of the steel hard?

5. Describe the reactions in the blast furnace.

6. Explain why carbon can reduce iron oxide, but does not affect calcium oxide.

7. Why must ships and bridges be inspected and painted frequently?

Completion Questions

Write the word, from the list below, that will correctly complete the following statements.

cast iron	taconite	flotation
ore	tempering	

1. An _____ is a natural mixture of minerals from which a valuable metal is extracted.

2. The high-temperature structure of a high-carbon steel is preserved at low temperatures by _____ .

3. A form of iron that cannot be welded is _____ .

4. A method used to separate the mineral from the gangue is _____ .

5. A second-rate ore of iron is _____ .

Find the Facts

Locate the sections in Chapter 9 that support each answer to each completion question.

True or False

If the statement is true, mark it *T*. If the statement is false, change the indicated word(s), using the list below, to make each statement true.

pig iron	flux	tempering
metals	steel	flotation

1. Ores are mixtures of *minerals.*

2. Most elements in the universe are *nonmetals.*

3. Aluminum is more active than *copper.*

4. The mineral is separated from its gangue by *heating.*

5. Another name for cast iron is *steel.*

6. Slag is formed in the reaction of the gangue with the *mineral.*

7. Rust in our drinking water comes from *cast-iron* water mains under the city streets.

8. *Cast iron* is an alloy of iron and carbon.

9. The highest quality steel is *electric-furnace* steel.

10. *Case hardening* is the process of heating a high-carbon steel, then quickly freezing the high-temperature structure.

Multiple-Choice

1. All of the following are ores of iron except:
   a. hematite
   b. magnetite
   c. taconite
   d. bauxite

2. All of the following metals are more active than hydrogen except:
   a. aluminum
   b. iron
   c. silver
   d. lead

3. Zinc will replace the metals in each of the following compounds except:

   a. $CuSO_4$ b. NaCl   c. $FeCl_3$ d. H_2O

4. The following steps are taken in the smelting of iron except:

   a. mining the ore
   b. reduction of the oxide
   c. changing the mineral to an oxide
   d. flotation

5. Most steel made today uses the following process:

   a. Bessemer process
   b. electric-furnace process
   c. open-hearth process
   d. instant process

Matching

Use Table 9-1 to find the ore in column B that relates most closely to the metal in column A.

COLUMN A	COLUMN B
_____ ore of iron	1. bauxite
_____ ore of copper	2. galena
_____ ore of uranium	3. taconite
_____ ore of aluminum	4. sphalerite
_____ ore of lead	5. malachite
_____ ore of titanium	6. uraninite
_____ ore of zinc	7. ilmenite

Part II

9-10   How Is Lead Extracted from the Earth?

Lead was used extensively by the Romans to make their water pipes. The Latin word for lead is **plumbum**. The chemical symbol for this element comes from Latin as **Pb**. Today people who work with pipes are called plumbers. Of course the Romans did not know that lead is more active than hydrogen (Table 9-2). As a result, lead dissolves slightly in water by replacing some of its hydrogen. Lead is a very poisonous metal. It is dangerous to drink water that flows through soldered pipes. The Romans suffered from lead poisoning. It causes the loss of intelligence and may result in madness. It is very likely that the emperors Nero and Caligula, who are remembered for their madness, may have been suffering the effects of lead poisoning. Cups and all cookware must never be allowed to become contaminated with lead. Some gasoline has lead added to it to improve engine performance. Unfortunately a gaseous lead compound ($PbBr_4$) is formed and is expelled from the tailpipe of the automobile. Breathing these leaded gases is also poisonous. Leaded gasoline will eventually be illegal. Lead is one of the few metals that the human body does not need even in trace quantities. All amounts of lead, no matter how small, are poisonous.

Galena is the ore from which our lead is obtained. The lead in galena is found as lead sulfide (PbS). After the ore is mined, the mineral is concentrated by the flotation process. The next task is to convert the sulfide into the oxide. The lead sulfide is **roasted** in a **reverberatory furnace** with excess air. The sulfur burns off as sulfur dioxide and lead oxide is produced. The heating continues until two thirds of the sulfide has been changed into the oxide. The air is then shut off and the remaining lead sulfide reduces the lead oxide to the free metal.

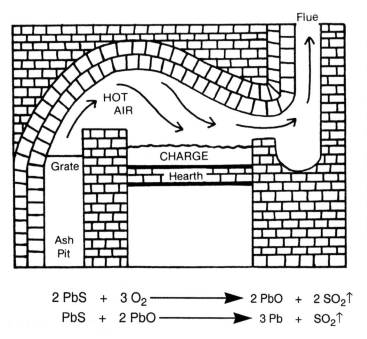

FIGURE 9-11

REVERBERATORY FURNACE

The lead obtained from the reverberatory furnace is called **pig lead**.

A reverberatory furnace reflects the hot flames and hot gases from its roof to the charge. This reflection echoes from the walls and roof, hence the name **reverberatory furnace**.

$$2 \, PbS + 3 \, O_2 \longrightarrow 2 \, PbO + 2 \, SO_2\uparrow$$
$$PbS + 2 \, PbO \longrightarrow 3 \, Pb + SO_2\uparrow$$

9-11 How Is Copper Extracted from the Earth?

Copper is one of the earliest metals used by humans. Tools, weapons, and utensils made of copper were better than those fashioned from stone. The introduction of a new and improved technology resulted in the Bronze Age, replacing the technology of the Stone Age. Bronze was another name for copper. Today bronze is an alloy of copper and tin.

Since copper is less active than hydrogen (see Table 9-2), it was found in its native state. Very little free copper can be found today because its supply was rapidly exhausted. When it is found free, the metal is simply heated in a furnace. The copper is melted and allowed to run off, leaving most of its impurities behind.

A major step in the extraction of any metal is to convert the mineral into its metallic oxide. The sulfide ores of copper are converted into its oxide by roasting similarly to that of galena (lead sulfide), described in section 9-10.

A common ore of copper is malachite, $CuCO_3$. The copper is extracted and refined in the following way. First the mineral is concentrated by flotation. The carbonate is then **calcined**, or heated in a furnace to convert the carbonate into its oxide.

Table 9-5

$$CuCO_3 \xrightarrow{\quad HEAT \quad} CuO \; + \; CO_2 \uparrow$$

Carbonate ores are decomposed into oxides.

$$CuO \; + \; C \xrightarrow{\quad HEAT \quad} Cu \; + \; CO \uparrow$$

Copper oxide is reduced with coke (carbon) in a furnace.

The oxide is then reduced to impure copper, called **matte**. The impure copper matte is then put into a Bessemer furnace, and air is forced through the molten mass. As burnable impurities are removed (sulfur and phosphorous), the molten mass thickens. When the process ends, blisters of air are trapped in the plastic mass. This product is called **blister copper**. It is about 99% pure copper. The 1% impurity consists of gold, silver, lead, and zinc. The copper must now be refined electrolytically.

9-12 How Is Copper Electrolytically Refined?

Plates of blister copper are immersed in an acidic solution of copper sulfate. This solution consists of ions [Cu^{+2} and $(SO_4)^{-2}$] in water (see section 5-6). The blister copper becomes the positive electrode. At that electrode, the copper atoms lose two electrons and become ions, which go into solution. The negative electrode attracts the positive copper ions in the solution. They gain electrons and are deposited as pure copper. The gold and silver impurities in the blister copper sink to the bottom of the tank as a sludge.

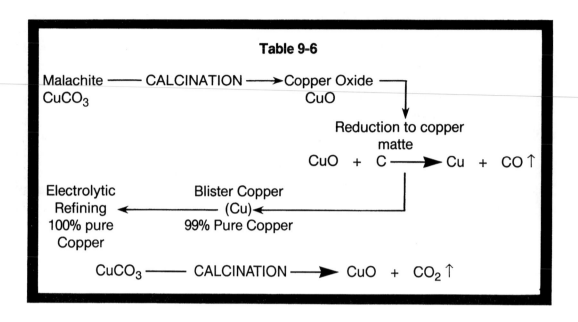

Table 9-6

Malachite ——— CALCINATION ———→ Copper Oxide
$CuCO_3$ CuO

Reduction to copper
matte

$$CuO + C \longrightarrow Cu + CO \uparrow$$

Electrolytic   Blister Copper
Refining   (Cu)
100% pure   99% Pure Copper
Copper

$$CuCO_3 \longrightarrow CALCINATION \longrightarrow CuO + CO_2 \uparrow$$

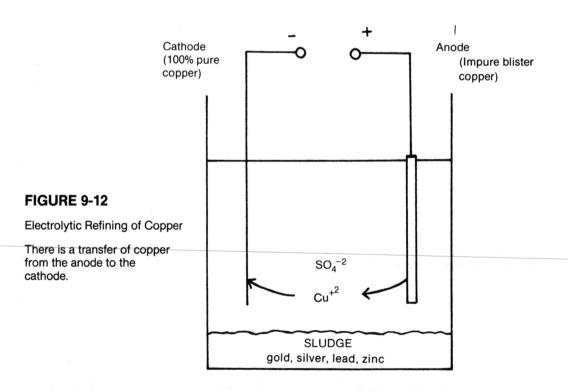

FIGURE 9-12

Electrolytic Refining of Copper

There is a transfer of copper from the anode to the cathode.

Cathode
(100% pure copper)

Anode
(Impure blister copper)

SO_4^{-2}

Cu^{+2}

SLUDGE
gold, silver, lead, zinc

$$CuO + C \longrightarrow Cu + CO \uparrow$$

$$CuCO_3 \longrightarrow CALCINATION \longrightarrow CuO + CO_2 \uparrow$$

Table 9-7

Positive Electrode Reaction $Cu^0 - 2e \longrightarrow Cu^{++}$

Negative Electrode Reaction $Cu^{++} + 2e \longrightarrow Cu^0$

Metallic copper ionizes from the positive electrode,
and deposits on the negative electrode.

9-13 How Did Aluminum, Our Most Abundant Metal, Become Available?

The tombs of the pharaohs were laden with gold. Gold adorns the ancient temples. Gold is the precious metal of royalty. Before 1825, no one knew about aluminum, our most abundant metal, but everyone knew about gold, which is one of our rarest metals. Why was aluminum unavailable? Aluminum is found in clay and other construction materials (see Chapter 8). Our best ore of aluminum is **bauxite**, which contains the mineral alumina (Al_2O_3). The problem with aluminum is its high activity (see Table 9-2: "Activity Series of Metals"). In order to free aluminum from its compounds, a more active metal than aluminum is needed. Aluminum was discovered in 1825 by the Danish scientist Hans Christian Oersted, who replaced the aluminum in aluminum chloride with potassium. Aluminum prepared in this way costs more than gold!

Table 9-8

$3 K^0 + AlCl_3 \longrightarrow 3 KCl + Al^0$

Aluminum can be liberated from its compounds by a more active metal.

$Carbon + Al_2O_3 \longrightarrow No\ Reaction$

Carbon is not active enough to replace aluminum from its oxide.

In 1885, a young student at Oberlin College heard his chemistry professor say that a fortune awaited the man who discovered an inexpensive way to obtain aluminum. The student, Charles Martin Hall, found the answer to the problem one year later. At the age of 22, he became a rich man. When he died in 1914, he left $15 million dollars to his college.

9-14 How Is Aluminum Liberated from the Earth?

After many attempts and failures, Hall decided to dissolve **alumina** (Al_2O_3) in molten cryolite. Cryolite is a rock that is easily melted. On passing an electric current through the dissolved alumina, aluminum is obtained.

FIGURE 9-13

Hall Process

$$Al_2O_3 \xrightarrow[\text{cryolite}]{\text{molten}} 2\,Al^{+++} + 3\,O^{=}$$

Negative Electrode Reaction

$$Al^{+++} + 3e \longrightarrow Al^0$$

Positive (carbon) Electrode Reactions

$$O^{=} - 2e \longrightarrow O^0$$

$$C + O \longrightarrow CO\uparrow$$

Atomic Oxygen, released at the positive carbon electrode by the electrolysis of Al_2O_3, burns the electrode to form carbon monoxide.

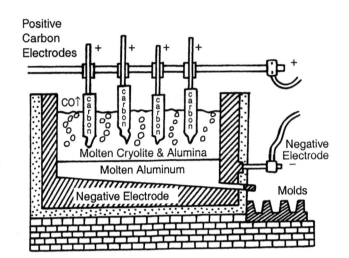

Aluminum oxide ionizes in molten cryolite. The oxide ions are attracted to the positive carbon (graphite) electrodes. The oxide ions give up their two electrons to the postive carbon electrode and become free oxygen. The oxygen burns the electrode to form a mixture of carbon monoxide and carbon dioxide. At the same time, aluminum ions pick up electrons from the negative electrode, and are deposited as molten aluminum. The molten metal is then run off into molds. Additional aluminum oxide is added from time to time to replenish the ions in the cryolite solution. Eventually, the metal obtained is 99.99% pure. Since the process requires large amounts of electricity, aluminum plants are located where there are abundant sources of electricity, such as in the Tennessee Valley.

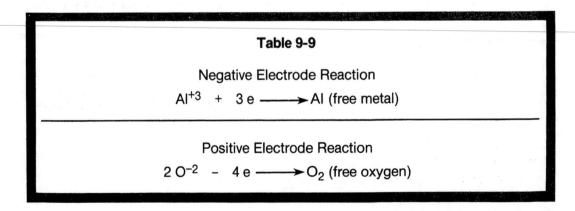

Table 9-9

Negative Electrode Reaction

$$Al^{+3} + 3\,e \longrightarrow Al \text{ (free metal)}$$

Positive Electrode Reaction

$$2\,O^{-2} - 4\,e \longrightarrow O_2 \text{ (free oxygen)}$$

9-15   What Is the Metal of Opportunity Today?

In 1885, aluminum was the metal of opportunity. Today, titanium is the metal of the future. Once again, a fortune awaits the person who develops an inexpensive way to obtain titanium. It is a metal that has great strength, resists corrosion, is usable at high temperatures, and is light in weight. Titanium, and its alloys are needed to line rocket and jet engines. Pure titanium is difficult to obtain. Titanium is the tenth most abundant element on the surface of our planet (see Table 8-2). Titanium is found in **rutile** as titanium dioxide (TiO_2), and in **ilmenite** as $FeTiO_3$.

9-16   How Is Titanium Extracted from Its Ores?

Titanium dioxide is such a stable compound that its molecules are found on cool, red giant stars. Carbon cannot separate the titanium from its oxygen. The process used today is the **Kroll process**. Chlorine gas is passed over a heated mixture of titanium dioxide, forming titanium tetrachloride.

$$TiO_2 \; + \; 2\,Cl_2 \longrightarrow TiCl_4 \text{ (liquid)} \; + \; \ldots$$

The titanium tetrachloride is then immersed in molten magnesium in an atmosphere of argon or helium. If oxygen were present, the magnesium would burn (combine with the oxygen) explosively. The more active magnesium replaces titanium in the compound.

$$TiCl_4 \; + \; 2\,Mg \longrightarrow 2\,MgCl_2 \; + \; Ti$$

The titanium obtained is called **sponge titanium**. It is impure. In order to obtain the pure metal, the impure sponge titanium is sublimated (see section 2-6) and then collected by crystallization.

The need for copper, aluminum, and titanium will continue to grow as the space age emerges. The major problems with the extraction of these metals are the great amounts of energy it requires, and the pollution it produces. The key to the problem may lie in new sources of pollution-free energy.

Now You Know

1. The chemical symbol for lead comes from its Latin word plumbum (Pb).

2. Sulfide ores must be changed into oxides by roasting.

3. Lead is obtained from galena by incomplete roasting.

4. Lead is slightly soluble in water, and is poisonous.

5. The chief copper ores are sulfides, oxides, and carbonates.

6. The sulfide ores of copper are first roasted to the oxide, then reduced to an impure copper matte.

7. The carbonate ores of copper must be calcined to the oxide, then reduced to an impure copper matte.

8. Copper ultimately must be electrolytically refined to be made 100% pure. This purity is needed for use in electrical wiring.

9. The chief ore of aluminum is bauxite. It contains the mineral alumina: Al_2O_3.

10. Aluminum is prepared by the Hall process.
    a. Alumina is dissolved in molten cryolite (a natural mineral), where the alumina ionizes.

$$Al_2O_3 \longrightarrow 2\ Al^{+3}\ +\ 3\ O^{-2}$$

    b. The alumina-cryolite solution is electrolyzed.

11. Titanium is the metal of the space age.

12. Titanium is very expensive and difficult to obtain in its pure state.

13. Pure titanium is obtained by the sublimation and then crystallization of impure sponge titanium.

New Words

alumina	Aluminum oxide: Al_2O_3
blister copper	The copper that results from the further reduction of the copper matte. It is 98% pure copper.
bronze	An alloy of copper and tin.

calcine	The heating of a carbonate to produce an oxide.
copper matte	An impure copper that is the product of the reduction of copper oxide.
cryolite	The mineral that is melted and used as the solvent for alumina in the Hall process.
reverberatory furnace	A furnace that reflects the hot flames and gases from the furnace roof to the charge.
roasting	The burning off of a sulfide ore to produce a metallic oxide.

Reading Power

For each of the following questions, select one answer that seems most correct.

1. The main idea of section 9-10 is:
   a. The Latin word for lead is plumbum.
   b. The Romans suffered from lead poisoning.
   c. Galena is roasted, then reduced in a reverberatory furnace.
   d. Lead is poisonous.

2. The main idea of section 9-11 is:
   a. The Bronze or Copper Age followed the Stone Age because copper was easily available.
   b. All carbonate ores must be calcined before they can be reduced.
   c. The smelting of copper involves many steps for its completion.
   d. Blister copper is 98   pure copper.

3. The main idea of section 9-12 is:
   a. Copper is refined with electricity.
   b. Blister copper is hung as the positive electrode.
   c. Pure copper deposits on the negative electrode.
   d. A sludge of gold and silver sinks to the bottom of the tank and is later recovered.

4. The main idea of section 9-13 is:
   a. Gold is a rare metal.
   b. Aluminum is obtained by the Hall process.
   c. Carbon cannot reduce aluminum oxide.
   d. Potassium can reduce aluminum oxide.

5. The main idea of section 9-14 is:
   a. Aluminum is obtained by electrolysis.
   b. Molten cryolite dissolves alumina.
   c. The Hall process requires cryolite.
   d. Carbon electrodes must be replaced in the Hall process because they are burned off to carbon dioxide and carbon monoxide by the released oxygen.

6. The main idea of section 9-15 is:

    a. Titanium is the space-age metal of opportunity today.
    b. Titanium is an abundant metal.
    c. Titanium is strong, it resists corrosion, is light in weight, and functions well under high temperatures.
    d. Rutile and ilmenite are ores of titanium.

7. The main idea of section 9-16 is:

    a. Titanium is an active metal.
    b. Pure titanium is obtained by the sublimation of sponge titanium.
    c. Sponge titanium is impure.
    d. Titanium is extracted from its minerals by the Kroll process.

Mind Expanders

1. Why did the Iron Age come after the Bronze Age?

2. Solder is an alloy of tin (33%) and lead (67%). Why should solder be avoided when fixing water pipes?

3. Describe the steps in the metallurgy of each of the following metals.
    a. lead   b. copper   c. aluminum   d. titanium

4. Why is lead from the reverberatory furnace called "pig lead"?

5. Why must copper be electrolytically refined?

6. Why are there no traces of aluminum objects from ancient Rome?

7. Why is titanium considered a space-age metal?

Completion Questions

Write the word(s) from the list below that will correctly complete the following statements.

bronze	bauxite	roasted	titanium	calcined

1. Sulfide ores must be _____ to be converted into an oxide.

2. Carbonate ores must be _____ to be converted into an oxide.

3. Ores of _____ cannot be reduced by carbon.

4. _____ is an alloy of copper and tin.

5. Alumina is the mineral found in _____ .

Find the Facts

Locate the section in Chapter 9 that supports each answer to the completion questions.

True or False

If the statement is true, mark it *T*. If the statement is false, change the indicated word(s), using the list below, to make the statement true.

aluminum	copper	plumbum
reverberatory furnace		titanium

1. The *Bessemer converter* reflects the hot gases from the roof of the furnace onto the charge.

2. Titanium is obtained by *sublimation*.

3. The most abundant metal on our planet is *copper*.

4. An element with a high melting point, great strength, and resistant to corrosion is *aluminum*.

5. Titanium is obtained by the *Kroll process*.

Multiple-Choice

1. A metal purified by sublimation is:
   a. copper   b. aluminum   c. titanium   d. lead

2. Malachite is the chief ore of:
   a. copper   b. aluminum   c. lead   d. titanium

3. The following mineral compound must undergo roasting:
   a. TiO_2 b. Al_2O_3 c. $CuCO_3$ d. CuS

4. The following mineral compound must be calcined:

    a. TiO_2 b. Al_2O_3 c. $CuCO_3$ d. CuS

5. A metal that must be electrolytically refined is:

    a. lead   b. copper   c. titanium   d. iron

Matching

Write the number in column B that relates to the words in column A in the spaces indicated.

COLUMN A		COLUMN B
_____ roasting	1.	A furnace that reflects hot gases.
_____ calcine	2.	Aluminum oxide.
_____ blister copper	3.	Converting a carbonate to an oxide.
_____ alumina	4.	About 98% pure copper.
_____ reverberatory furnace	5.	Conversion of a sulfide ore into an oxide.

Related Science Careers

High School Diploma	
blacksmith, coremaker, electroplater, foundry worker, jeweler, sheet-metal worker, smelter, welder	
2-Year Degree	**4-Year or Graduate Degree**
metallurgical technologist	metallurgical engineer
lab technician	chemical engineer
	chemist

Chapter 10

Why Do We Study Carbon and Its Compounds?

Instructional Objectives

After completing this chapter, you will be able to:

1. Define alcohol, aldehyde, aliphatic hydrocarbons, alkanes, alkenes, alkynes, aromatic hydrocarbons, biochemistry, denatured, ester, fermentation, hydrocarbon, isomer, ketone, organic acid, organic chemistry, oxidation, and reduction.

2. Recognize the uniqueness of the element, carbon.

3. Distinguish between organic chemistry and biochemistry.

4. Explain how carbon atoms combine.

5. Classify and compare hydrocarbon families.

6. Recognize isomers.

7. Recognize addition reactions.

8. Compute the formulas of aliphatic hydrocarbons from their general formulas.

9. Recognize alcohols, aldehydes, ketones, acids, and esters by their functional groups.

10. Outline the products of oxidation from the hydrocarbon to the acid.

Chapter 10 Contents

Part I

Introduction   **235**

10-1   How Do Carbon Atoms Combine?   **236**

10-2   How Did Organic Chemistry Originate?   **236**

10-3   What Are the Main Groups of Hydrocarbons?   **237**

10-4   How Are the Alkanes Different from Other Families?   **237**

10-5   What Are Isomers?   **239**

10-6   How Are the Physical Properties of Aliphatic Compounds Affected by Molecular Weight?   **239**

10-7   How Are the Alkenes Different from Other Families?   **241**

10-8   How Are the Alkynes Different from Other Families?   **242**

10-9   How Can We Determine the Formulas for the Aliphatic Hydrocarbons?   **243**

Written Exercises   **245**

Part II

10-10   How Are Alcohols Made?   **250**

10-11   How Are Alcohols Oxidized?   **252**

10-12   How Are Carbohydrates Fermented?   **253**

10-13   What Is an Organic Acid?   **256**

10-14   How Do Acids React with Alcohols?   **257**

Written Exercises   **260**

Chapter 10

Why Do We Study Carbon and Its Compounds?

Part I

Introduction

Only one out of every thousand atoms on the surface of the earth is carbon. You may wonder why there is a whole chapter in this book devoted to such a rare element. Carbon is a most unusual element. It forms the basis for all living things. Carbon is found in all foods. Carbon is different from all other elements because it is able to bond to itself in almost endless chains, shapes, and designs. Carbon compounds are able to change their molecular arrangement. This ability allows living things to change and adapt to new conditions. There are more different carbon compounds than all noncarbon compounds put together. The branch of chemistry that deals with carbon compounds is called **organic chemistry**.

Table 10-1

Straight-Chain Molecules

Branched-Chain Molecules

ring or cyclic molecules

Phenanthrene

Carbon is a rare element, but it forms more different compounds than all other elements put together.

10-1   How Do Carbon Atoms Combine?

The carbon atom has four electrons in its valence shell. They are capable of sharing electrons with each other to form covalent bonds (see Chapter 4).

Table 10-2

Every carbon atom must have four bonds in every compound.

$$\cdot \dot{C} \cdot \; + \; \cdot \dot{C} \cdot \; + \; \cdot \dot{C} \cdot \; + \; \cdot \dot{C} \cdot \longrightarrow \cdot \dot{C} : \dot{C} : \dot{C} : \dot{C} \cdot$$

When carbon atoms share one pair of electrons, they are bonded by a single covalent bond.

$$-\overset{|}{\underset{|}{C}}-\overset{|}{\underset{|}{C}}-\overset{|}{\underset{|}{C}}-\overset{|}{\underset{|}{C}}-$$
The dash stands for a pair of shared electrons (a covalent bond).

$$:C::\dot{C}.$$
When carbon atoms share two pairs of electrons, they are bonded by a double covalent bond.

$$\underset{}{\supset} C = C \underset{}{\subset}$$
Two dashes stand for two pairs of shared electrons. Two dashes represent a double covalent bond.

$$\cdot C:::C\cdot$$
When carbon atoms share three pairs of electrons, they are bonded with a triple covalent bond. This is represented by three dashes.

$$-C \equiv C-$$

Two carbon atoms are not able to form four covalent bonds with each other.

10-2   How Did Organic Chemistry Originate?

Before 1828, scientists thought that organic compounds could only be made by plants and animals. In that year, Friedrich Wöhler, made urea from ammonium cyanate.

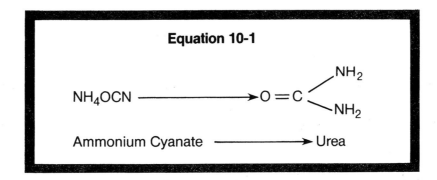

Until this time, it had been believed that urea could be made only by living animals. This discovery led to the establishment of modern organic chemistry as the chemistry of carbon compounds. The study of compounds associated with living things is called **biochemistry**. Because of his discovery, Friedrich Wöhler is now considered to be the father of organic chemistry.

10-3   What Are the Main Groups of Hydrocarbons?

Hydrocarbons are compounds of carbon and hydrogen. They are classified into two large groups: the **aliphatic hydrocarbons**, and the **aromatic hydrocarbons**. Aliphatic is an Arabic word that means "fatty," and aromatic means "fragrant." To the organic chemist, an aromatic compound is one that contains a particularly stable ring of carbon atoms with special properties. An aliphatic hydrocarbon is one that isn't aromatic. There are three major groups of aliphatic compounds. They are the **alkanes**, the **alkenes**, and the **alkynes**.

10-4   How Are the Alkanes Different from Other Families?

The alkane hydrocarbons are made of chains of carbon atoms. They only have single bonds between their carbon atoms. The first member of this "family" or series of compounds is methane (CH_4). The alkanes are often called the "methane series." Families are named after the first member of the series.

There are two hydrogens for each carbon atom, and two more for the ends of the chain. The general formula for any alkane is $C_nH_{(2n+2)}$. Just as all members of your family have the same last name, the members of the alkane series also have the same family name. They are the "**ane**" family. All alkanes can be identified because the name of each alkane compound ends in -*ane*. Other names for methane are natural gas and marsh gas. Methane and other alkanes are found naturally in oil wells. Methane is produced in swamps by decaying vegetation under the mud and in landfills by decaying garbage (see section 13-9e). It is a fuel used in your stove at home.

Table 10-3

The Hydrocarbons

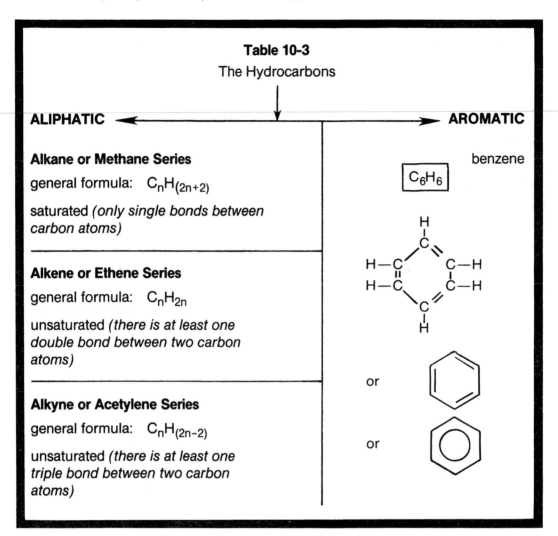

ALIPHATIC ⟵————————————⟶ **AROMATIC**

Alkane or Methane Series

general formula: $C_nH_{(2n+2)}$

saturated *(only single bonds between carbon atoms)*

Alkene or Ethene Series

general formula: C_nH_{2n}

unsaturated *(there is at least one double bond between two carbon atoms)*

Alkyne or Acetylene Series

general formula: $C_nH_{(2n-2)}$

unsaturated *(there is at least one triple bond between two carbon atoms)*

benzene

C_6H_6

or

or

Table 10-4

CH_4	C_2H_6	C_3H_8	C_4H_{10}
methane	ethane	propane	butane

pentane

C_5H_{12}

Equation 10-2

$$CH_4 + 2O_2 \longrightarrow CO_2 + 2H_2O + HEAT$$

10-5 What Are Isomers?

The alkane that has four carbon atoms and ten hydrogen atoms can be arranged in two different ways.

Table 10-5

n-butane

(n means "normal" or straight chain)

isobutane

Butane and isobutane are two different compounds that have the same formula: C_4H_{10}. Compounds with the same formula but different molecular structure are called **isomers**. "Iso-" is a Greek word that means same or equal. "Mer" means unit or formula. Isomers are different compounds, even though they have the same formula. Each isomer has its own properties. The molecular arrangement is the most important feature in determining the identity of the material.

10-6 How Are the Physical Properties of Aliphatic Compounds Affected by Molecular Weight?

As the molecules of the alkane series grow larger and heavier, their boiling point increases and the freezing point also rises. Small, light molecules exist as gases at room temperature. As the molecules get heavier, the alkane changes to a liquid and eventually becomes a solid. The information in Table 10-7 is for "normal" or straight-chain alkanes. Isomers will each have different boiling points and freezing points.

Table 10-6

There are three isomers of pentane. C_5H_{12}

Can you draw the five isomers of hexane (C_6H_{14})?

Table 10-7   The Alkane or Methane Series

Name	Formula	Molecular Weight	Boiling Point ° C	Melting Point ° C	Physical State
methane	CH_4	16	– 161.5	– 190	gas
ethane	C_2H_6	30	–88.3	– 184	gas
propane	C_3H_8	44	– 42.2	– 172	gas
butane	C_4H_{10}	58	0	– 135	gas
pentane	C_5H_{12}	72	36.2	– 130	liquid
hexane	C_6H_{14}	86	69	– 94.3	liquid
heptane	C_7H_{16}	100	98.4	– 90.5	liquid
octane	C_8H_{18}	114	114	– 56.5	liquid
nonane	C_9H_{20}	128	150.8	– 53.7	liquid
decane	$C_{10}H_{22}$	142	174	– 31	liquid
cetane (hexadecane)	$C_{16}H_{34}$	226	287.5	+ 20	solid

10-7 How Are the Alkenes Different from Other Families?

There is much scientific truth to the statement that one rotten apple will spoil the rest. Ethene is a gas that is formed by rotting fruits. This odorless, invisible gas will speed the ripening (and rotting) of fresh fruit, and it is often used to force plants to flower.

Ethene (C_2H_4) is the first member of the alkene series. They are straight-chain hydrocarbons that have at least one double bond between two carbon atoms. This group of hydrocarbons is also known as the ethene series.

Table 10-8

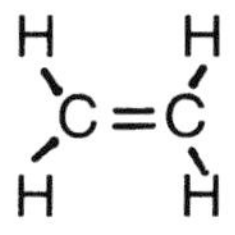 Ethene	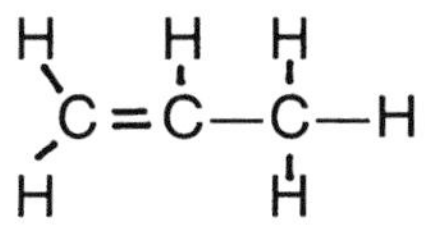 Propene	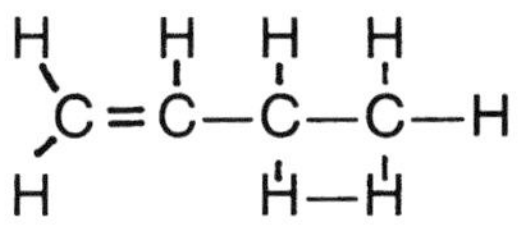 1–Butene
C_2H_4	C_3H_6	C_4H_8

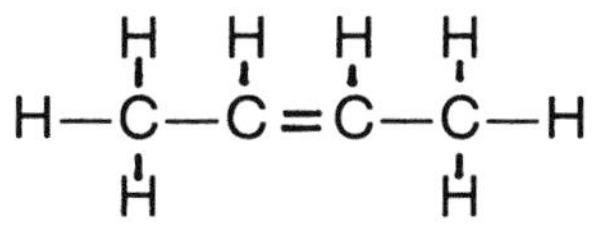

2–Butene

One isomer of butene has its double bond after the first carbon atom. Another isomer has its double bond after the second carbon atom.

All alkenes end in *-ene*. They can be recognized by these letters at the end of the names of their compounds. The introduction of the double bond removes two hydrogen atoms from the molecule. These compounds will have two hydrogen atoms less than the alkanes. The general formula for the alkenes is C_nH_{2n}. The alkenes are unsaturated because of the double bond. Other elements and radicals can add to the double bond.

Table 10-9

Addition Reactions at the Double Bond

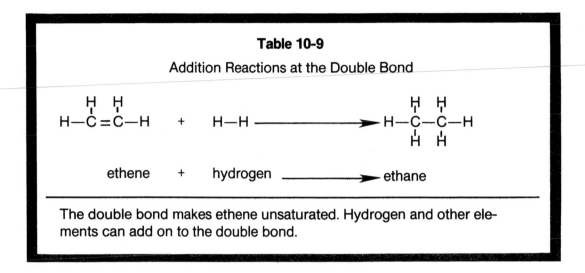

ethene   +   hydrogen ——————→ ethane

The double bond makes ethene unsaturated. Hydrogen and other elements can add on to the double bond.

10-8   How Are the Alkynes Different from Other Families?

Acetylene is a colorless gas used for early automobile headlights. It was produced by dripping water onto calcium carbide. The gas burned with a bright flame. Today, some campers use acetylene lamps. The gas is also used for underwater welding in oxyacetylene torches.

Equation 10-3

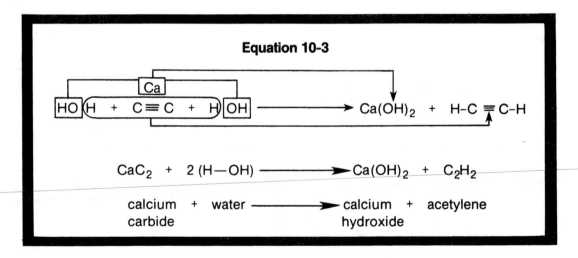

$$CaC_2 + 2(H-OH) \longrightarrow Ca(OH)_2 + C_2H_2$$

calcium   +   water ——————→ calcium   +   acetylene
carbide   hydroxide

Acetylene (or ethyne) is the first member of the alkyne series. This series contains **aliphatic** compounds that have a triple bond between two carbon atoms. This group of compounds is also known as the ethyne series, or acetylene series. All alkynes end in -*yne*. The introduction of the triple bond reduces the number of hydrogen atoms in the molecule to two less than in the alkenes. The general formula for the alkyne series is $C_nH_{(2n-2)}$. The alkynes are more unsaturated than the alkenes because of their triple bond. Hydrogen and other substances can add to this bond.

Table 10-10

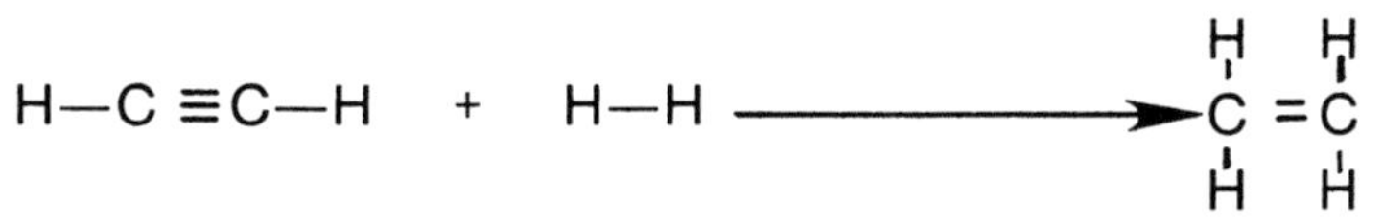

$$H-C \equiv C-H \quad + \quad H-H \longrightarrow \underset{\underset{H}{|}}{\overset{\overset{H}{|}}{C}} = \underset{\underset{H}{|}}{\overset{\overset{H}{|}}{C}}$$

ethyne   +   hydrogen ————————▶ ethene
*(acetylene)

---

*Acetylene, which ends in "ene," is the common name for C_2H_2.
Ethyne is the name internationally agreed upon by chemists.

Table 10-11

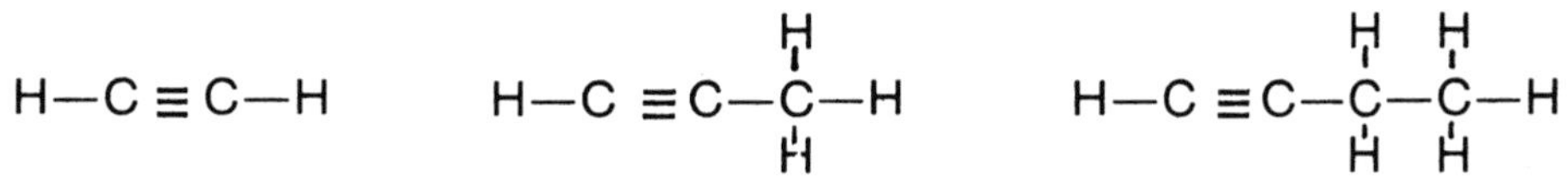

$$H-C \equiv C-H \qquad H-C \equiv C-\underset{\underset{H}{|}}{\overset{\overset{H}{|}}{C}}-H \qquad H-C \equiv C-\underset{\underset{H}{|}}{\overset{\overset{H}{|}}{C}}-\underset{\underset{H}{|}}{\overset{\overset{H}{|}}{C}}-H$$

$CH \equiv CH$ $CH \equiv C-CH_3$ $CH \equiv C-CH_2-CH_3$

ethyne   propyne   butyne

10-9   How Can We Determine the Formulas for the Aliphatic Hydrocarbons?

The general expression for all alkanes is $C_nH_{(2n+2)}$ (see section 10-4). This allows us to determine the formula for all family members. In hexane, for example, we have six carbon atoms in the molecule (n = 6). To determine the number of hydrogen atoms in the molecule we must substitute "6" for "n." 2n + 2 = 2(6) + 2 or 12 + 2 = 14. The formula for hexane is C_6H_{14}. Determine the formulas for other aliphatics in the exercises that follow.

Formula Determination Problems

The general expression for all alkanes is $C_nH_{(2n + 2)}$.
Complete the formulas for each of the following compounds.

$C_{21}H_u$	$C_{33}H_v$	$C_{55}H_w$
C_xH_{78}	C_yH_{66}	C_zH_{54}

Determine the correct alkene and alkyne formulas for each of the above.

Now You Know

1. Carbon is a rare element on the earth.

2. Carbon is different from all other elements because it can covalently bond to itself in almost endless chains and designs. As a result, there are more different carbon compounds than all noncarbon compounds put together.

3. Carbon compounds form the molecular basis for all living things.

4. Organic chemistry is the study of carbon compounds.

5. Carbon forms covalent bonds with itself and other atoms. Carbon may form single, double, or triple bonds with itself. Carbon cannot form four covalent bonds with itself.

6. Hydrocarbons are compounds of carbon and hydrogen.

7. There are two major groups of hydrocarbons: the aliphatics and the aromatics.

8. An aromatic compound is one that contains a particularly stable ring of carbon atoms with special properties. An aliphatic hydrocarbon is one that isn't aromatic.

9. The three major groups that compose the aliphatic hydrocarbons are the alkanes, alkenes, and alkynes.

10. The alkanes are also known as the methane series.

11. The alkanes are saturated hydrocarbons (they have only single bonds between their carbon atoms). The general formula for the alkanes is $C_nH_{(2n + 2)}$.

12. Members of the alkane series can be recognized by their family suffix "ane" at the end of the names of all alkane compounds.

13. Isomers are compounds that have the same formula, but different molecular arrangements.

14. The alkenes are aliphatic hydrocarbons that contain at least one double bond between carbon atoms. The general formula for all alkenes is C_nH_{2n}. They are considered unsaturated because of their double bond. All alkenes can be recognized by their "ene" ending.

15. The alkynes are aliphatics that contain a triple bond between carbon atoms. The general formula for the alkynes is $C_nH_{(2n-2)}$. They are unsaturated because of the triple bond. All alkyne compounds can be recognized because their names end in "yne."

New Words

aliphatic hydrocarbons	Hydrocarbons that are not aromatic.
alkanes	Aliphatic hydrocarbons in which there are only single bonds between carbon atoms.
alkenes	Aliphatic hydrocarbons in which there is at least one double bond between carbon atoms.
alkynes	Aliphatic hydrocarbons that have a triple bond between carbon atoms.
aromatic hydrocarbons	Cyclic or ring compounds of carbon that have special properties.
biochemistry	The study of compounds and reactions that are involved with living things.
hydrocarbons	Compounds of carbon and hydrogen.
isomer	Compounds with the same formula but different molecular structure or atomic arrangement.
organic chemistry	The study of carbon compounds.

Reading Power

For each of the following questions, select one answer that seems most correct.

1. The main idea of the introduction is:
   a. Carbon is a very rare element.
   b. Carbon is a very useful element.
   c. Carbon forms the molecular basis of all life.
   d. Organic chemistry deals with carbon compounds.

2. The main idea of section 10-1 is:

    a. Carbon can combine only with itself.
    b. Carbon forms ionic bonds with other atoms.
    c. Carbon can form single, double, or triple covalent bonds with itself and other elements.
    d. Every carbon atom must have four bonds in every compound.

3. The main idea of section 10-2 is:

    a. Organic chemistry is the chemistry involved with living things.
    b. Organic chemistry had to be redefined.
    c. Biochemistry is the branch of chemistry that deals with living things.
    d. Organic chemistry originated when Friedrich Wöhler made urea from ammonium cyanate.

4. The main idea of section 10-3 is:

    a. Hydrocarbons are compounds of carbon and hydrogen.
    b. Aliphatic hydrocarbons are straight-chain compounds.
    c. Aromatic hydrocarbons are ring compounds.
    d. Hydrocarbons are either aliphatic or aromatic.

5. The main idea of section 10-4 is:

    a. The alkanes are aliphatic hydrocarbons.
    b. There are only single bonds between carbon atoms of the alkanes.
    c. The alkanes have a general formula.
    d. All alkanes are hydrocarbons.

6. The main idea of section 10-5 is:

    a. Isomers are compounds that have the same formula (unit), but differ in their molecular structure.
    b. "Iso" is the Greek word for "same" or "equal."
    c. Some compounds have many isomers.
    d. All substances with the same formula are the same.

7. The main idea of section 10-6 is:

    a. The molecular weight does not affect the properties of a compound.
    b. Isomers of the same compound have the same molecular weight.
    c. The greater the molecular weight, the greater the number of atoms in the formula.
    d. Substances with greater molecular weights have higher boiling and melting points.

8. The main idea of section 10-7 is:

    a. All alkenes have at least one double bond between two carbon atoms.
    b. All alkene compounds are aliphatic.
    c. Alkenes are straight-chain hydrocarbons.
    d. The alkenes have many isomers.

9. The main idea of section 10-8 is:

    a. All alkynes are aliphatic hydrocarbons.

    b. All alkynes have a triple bond between carbon atoms.

    c. Alkynes are aromatic hydrocarbons.

    d. The alkynes are aliphatic hydrocarbons.

Mind Expanders

1. How can isomerism allow living things to adapt to change?

2. What is meant by the term "polyunsaturated" oils?

3. Why are there many more organic compounds than inorganic compounds?

4. Draw the isomers of octane C_8H_{18}.

Matching

Write the number in column B that relates to the word(s) or formulas in column A in the spaces indicated.

COLUMN A	COLUMN B
_____ $C_nH_{(2n+2)}$	1. pentyne
_____ C_5H_{10}	2. pentene
_____ C_6H_6	3. pentane
_____ C_5H_{12}	4. alkane
_____ C_5H_8	5. straight-chain hydrocarbon
_____ aliphatic	6. aromatic hydrocarbon

Table 10-12

Fill in the blank areas of the table below.

Name	Structural Formula
1-butene	
isobutane	
1-propyne	
pentane	
3-octene	

Multiple-Choice

1. All of the following are organic compounds except:

   a. C_2H_2 b. H_2O c. $CH_3—CH_3$ d. $CH_3—CH_2—CH_3$

2. All of the following compounds are aliphatic except:

   a. C_6H_{14} b. C_6H_{10} c. C_6H_6 d. C_6H_{12}

3. All of the following compounds are alkanes except:

   a. C_5H_{12} b. C_7H_{16} c. C_3H_6 d. $C_{12}H_{26}$

4. All of the following compounds are alkenes except:

   a. C_8H_{16} b. C_3H_6 c. $C_{15}H_{30}$ d. C_4H_{10}

5. All of the following are alkynes except:

   a. C_9H_{18} b. C_5H_8 c. $C_{13}H_{24}$ d. C_4H_6

6. Ethane, ethene, and ethyne are all similar in that they are:

   a. hydrocarbons   b. all unsaturated

   c. all saturated   d. aromatic

7. Which pair of compounds illustrates isomerism?

   a. $H-\overset{\displaystyle H}{\underset{\displaystyle H}{C}}-\overset{\displaystyle H}{\underset{\displaystyle H}{C}}-H$ and $\overset{\displaystyle H}{\underset{\displaystyle H}{C}}=\overset{\displaystyle H}{\underset{\displaystyle H}{C}}$

   b. $H-C\equiv C-\overset{\displaystyle H}{\underset{\displaystyle H}{C}}-H$ and $H-\overset{\displaystyle H}{\underset{\displaystyle H}{C}}-C\equiv C-H$

   c. $H-\overset{\displaystyle H}{\underset{\displaystyle H}{C}}-\overset{\displaystyle H}{\underset{\displaystyle CH_3}{C}}-\overset{\displaystyle H}{\underset{\displaystyle H}{C}}-H$ and $H-\overset{\displaystyle H}{\underset{\displaystyle H}{C}}-\overset{\displaystyle H}{\underset{\displaystyle H}{C}}-\overset{\displaystyle H}{\underset{\displaystyle H}{C}}-\overset{\displaystyle H}{\underset{\displaystyle H}{C}}-H$

   d. $H-\overset{\displaystyle H}{\underset{\displaystyle H}{C}}-Cl$ and $Cl-\overset{\displaystyle H}{\underset{\displaystyle H}{C}}-H$

8. What is the total number of carbon atoms in a molecule of pentyne?

   a. 1   b. 2   c. 3   d. 5

9. How many covalent bonds are there in a molecule of propane?

   a. 3   b. 9   c. 10   d. 12

10. The alkane series has the general formula:

    a. $C_nH_{(2n-2)}$ b. C_nH_{2n} c. $C_nH_{(2n+2)}$ d. $C_nH_{(2n-6)}$

Find the Facts

Locate the sections in Chapter 10 that support each answer to the multiple-choice questions. Copy the section number into your notebook near your answer.

Completion Questions

Complete each of the following statements using the formulas below. You may use the same formula more than once.

$$C_3H_8 \qquad\qquad CH_2 \qquad\qquad C_4H_6$$

1. A member of the alkyne series has the formula _____ .

2. The third member of the alkane series has the formula _____ .

3. The formula for a saturated hydrocarbon is _____ .

4. Each member of the alkane series differs from the next member by

    _____ .

True or False

Mark each of the following statements true (*T*) or false (*F*). If the statement is false, replace the indicated word(s), using the list below, to make the statement true.

addition	alkene series	biochemistry

1. C_6H_{12} represents a member of the *methane series*.

2. $C_nH_{(2n-2)}$ is the general formula for the *alkyne series*.

3. The reaction $C_2H_4 + H_2 \longrightarrow C_2H_6$ is an example of a(n) *combustion* reaction.

4. Benzene is an example of an *aromatic* hydrocarbon.

5. The chemistry of living things is called *organic chemistry*.

Part II

10-10   How Are Alcohols Made?

If aliphatic hydrocarbons are oxidized slowly, in controlled steps, oxygen is added to the hydrocarbon chain. A hydroxyl (—OH) group is formed. A hydrocarbon that has a hydroxyl group is called an **alcohol**. The hydroxyl (—OH) is called a **functional group** because many of the reactions that characterize alcohols involve this group. All alcohols can be recognized by the "ol" ending in their names.

Equation 10-4

$$H-\overset{\overset{\textstyle H}{|}}{\underset{\underset{\textstyle H}{|}}{C}}-H \ + \ \text{oxygen} \longrightarrow H-\overset{\overset{\textstyle H}{|}}{\underset{\underset{\textstyle H}{|}}{C}}-OH$$

methanol or
methyl alcohol
(wood alcohol)

Methyl alcohol is also known as wood alcohol because it can be obtained from wood. Wood alcohol is extremely poisonous. Breathing its vapors or drinking it will cause blindness or death.

Equation 10-5

$$H-\overset{\overset{\textstyle H}{|}}{\underset{\underset{\textstyle H}{|}}{C}}-\overset{\overset{\textstyle H}{|}}{\underset{\underset{\textstyle H}{|}}{C}}-\overset{\overset{\textstyle H}{|}}{\underset{\underset{\textstyle H}{|}}{C}}-H \ + \ \text{oxygen} \longrightarrow$$

propanol
or propyl
alcohol

isopropanol
or isopropyl
alcohol
(rubbing
alcohol)

Oxygen can be added to the end of the chain or into the middle of it. When propane is oxidized to alcohol, two isomers are formed.

Equation 10-6

$$*R-H \ + \ \text{oxygen} \longrightarrow R-OH$$

hydrocarbon   alcohol

*R—H is the general formula for any hydrocarbon, and R—OH is the general formula for any alcohol.

• The symbol "R" stands for any hydrocarbon radical. A hydrocarbon with one hydrogen removed from it is a radical.

Industrial alcohol is denatured (poisoned) to make it unfit to drink. The wood alcohol or other poisonous ingredients added to it cannot be removed.

10-11   How Are Alcohols Oxidized?

Oxidation in organic chemistry is the addition of oxygen, or the loss of hydrogen, from an organic molecule. Hydrogen can be removed from alcohol with the help of an **oxidizing agent**. If methanol is heated with copper oxide (CuO), it will dehydrogenate (oxidize) the alcohol.

Equation 10-7

$$H-\underset{\underset{H}{|}}{\overset{\overset{H}{|}}{C}}-OH \ + \ CuO \longrightarrow H-C\overset{\displaystyle O}{\underset{\displaystyle H}{\Big<}} \ + \ H_2O \ + \ Cu$$

methanol   formaldehyde
(methaldehyde)

The products of this reaction are water, copper, and formaldehyde. Formaldehyde is a poisonous fluid used to preserve biological materials. Another name for this material is methaldehyde. It is the simplest of a group of organic compounds known as **aldehydes**. The general formula for an aldehyde is

$$R-C\overset{\displaystyle O}{\underset{\displaystyle H}{\Big<}} \qquad or \qquad RCHO$$

R represents an organic radical.

If the alcohol group ($-OH$) is in the middle of the carbon chain, a ketone is formed.

Table 10-13

H—C—C—C—H ——————— (− 2H) ————►H—C—C—C—H

propanol
(rubbing alcohol)

dimethyl ketone
(acetone)

The general formula for a ketone is:

$$O$$
$$R—C—R'$$ or $$R—CO—R'.$$

R' may be a different organic radical from R.

The carbon-oxygen groups are polar, and attract water. As a result, sugar attracts water (hygroscopic) to itself. This is why sugar will "cake up" if exposed to air. Apples are rich in sugar. If an apple is placed in a closed container with bread, the sugar in the apple will speed the drying of the bread, as it absorbs the moisture from it. On the other hand, if apples are placed in a cake, the apples will hold the moisture and prevent the cake from drying out. Candy (especially candy containing honey) is often sticky because of the absorption of moisture.

10-12 How Are Carbohydrates Fermented?

Fermentation is the process by which carbohydrates (sugar and starch) are reduced to alcohol and carbon dioxide. It is the oldest chemical process known to civilization. Winemaking is recorded in the earliest records of ancient cultures. Wine was and still is used in religious ceremonies. Fermentation is also important in the baking process (section 11-3c). Yeast must be present for fermentation to take place.

Wines are made by fermenting fruits (grapes, apples, plums, berries, etc.). Fruit sugar will ferment until the alcohol concentration becomes about 15%. At this point, the yeast will die and fermentation will end. Hard liquors such as gin and bourbon have higher concentrations of alcohol. The added alcohol is obtained by distillation. Rum is made from fermented molasses and cane sugar. Vodka is made by the fermentation of potatoes. The proof number on the label of an alcoholic beverage is twice the percent of alcohol. For example, a 90-proof liquor is 45% alcohol.

FIGURE 10-1a

Fermentation: The oldest chemical process

1. Hand-picked grapes arriving at the winery in preparation for pressing.

FIGURE 10-1b

2. Fermentation takes place in stainless steel tanks under controlled conditions.

Courtesy: Brotherhood Winery; Washingtonville, New York

FIGURE 10-1c

3. Freshly fermented clear wine stored in casks.

FIGURE 10-1d

4. The winemaster in his laboratory. Chemistry eliminates the guesswork in winemaking.

Courtesy: Brotherhood Winery; Washingtonville, New York

Equation 10-8

fruit + yeast ———— fermentation ———→ ethyl + carbon
sugar   alcohol   dioxide

$$C_6H_{12}O_6 \xrightarrow{\text{fermentation}} 2\,C_2H_5OH + 2\,CO_2$$

fructose   ethanol
(fruit sugar)   (ethyl or grain
                                                   alcohol)

Ethanol is the least poisonous of all alcohols.

Equation 10-9

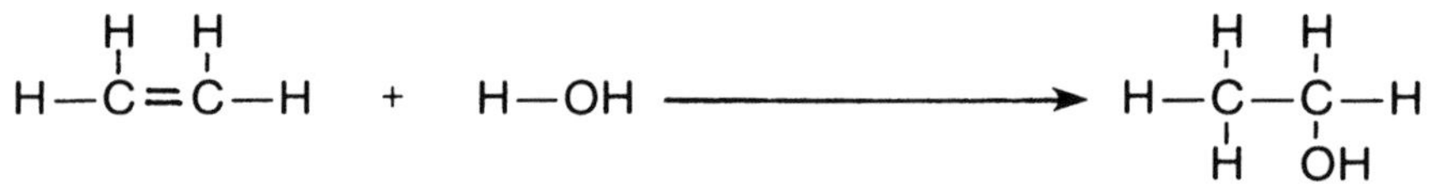

ethene + water ————————→ ethanol

Alcohol can be synthesized, without fermentation, from ethene and water. The law requires that all alcoholic beverages be produced from the fermentation of fruits, vegetables, and grains.

10-13   What Is an Organic Acid?

If the aldehyde continues to be oxidized, its molecule will gain additional oxygen. Eventually, an acid will form. The —COOH is the functional group that identifies all organic acids. It is called a **carboxyl group**. The hydrogen in the carboxyl group ionizes in water. All organic acids are weak because they are only partially ionized.

Equation 10-10

ethane   ——— oxygen ———→   ethanol

ethanol   ——— oxidation ———→   ethaldehyde

ethaldehyde   ——— oxidation ———→   ethanoic acid (acetic acid)

Oxidation is the gain of oxygen or the loss of hydrogen. Reduction is the opposite of oxidation.

10-14 How Do Acids React with Alcohols?

Acids react with alcohols to form a class of compounds called **esters**. Esters have many uses as artificial food odors and as polyester textiles. Water must be split out of the acid and alcohol molecules when esters form. Concentrated sulfuric acid is needed in the preparation of an ester. The sulfuric acid acts as a dehydrating agent. The general formula for an ester is R—COOR′

$$R-C\overset{O}{\underset{OR'}{\big\langle}}.$$

(R and R′ may be two different organic radicals.)

Equation 10-11

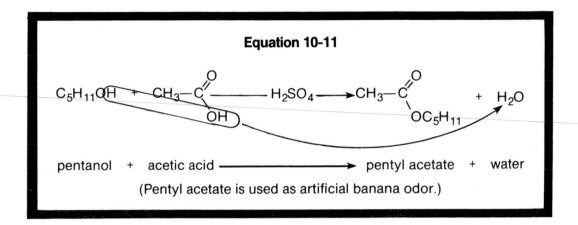

pentanol + acetic acid ⟶ pentyl acetate + water

(Pentyl acetate is used as artificial banana odor.)

You have been exposed to a brief survey of organic chemistry. There are many kinds of organic compounds that were not discussed. In the chapters that follow, you will explore applications of this vast branch of chemistry and how it has changed our lives.

Now You Know

1. The steps in hydrocarbon oxidation are as follows:

   hydrocarbon ⟶ alcohol ⟶ aldehyde ⟶ acid
   or ketone

2. Oxidation, in organic chemistry, is either the gain of oxygen, or the loss of hydrogen, by a hydrocarbon.

3. Reduction, in organic chemistry, is the gain of hydrogen or the loss of oxygen.

4. The functional group for all alcohols is —OH.

5. A functional group is a group of atoms in a molecule that is responsible for the properties of the compound. It characterizes and classifies the compound. For example, the —OH group classifies organic compounds as alcohols.

6. Alcohols can be oxidized to aldehydes or ketones.

7. Carbohydrates are aldehydes and ketones.

8. Carbohydrates are reduced to alcohol as a result of fermentation.

9. Carbohydrates and aldehydes can be oxidized to organic acids. The functional group of an organic acid is the carboxyl group: —COOH. The general formula for an organic acid is R—COOH.

10. Acids react with alcohols to form esters. A strong dehydrating agent is needed to split out the water formed in the reaction. Concentrated sulfuric acid is used for this purpose.

11. The general formula for an ester is R—COOR′.

12. Denatured alcohol is grain alcohol (ethanol) made for industrial use. It has poisons added to it that are most difficult to remove.

New Words

alcohol An organic compound that has the hydroxyl (—OH) group for its functional group.

aldehyde An organic compound that has $-\overset{O}{\underset{H}{\overset{\parallel}{C}}}$ or (—CHO) for its functional group.

denatured alcohol Industrial alcohol that is rendered poisonous by the addition of chemicals that cannot be easily removed from it.

ester An organic compound formed by the reaction between an organic acid and an alcohol. Its general formula is R—COOR′.

fermentation The reduction of a carbohydrate to an alcohol and carbon dioxide.

ketone An organic compound that has $-\overset{O}{\overset{\parallel}{C}}-$ or (—CO—) for its functional group.

organic acid An organic compound that has the carboxyl (—COOH) as its functional group.

oxidation The gain of oxygen or the loss of hydrogen by an organic compound.

reduction The gain of hydrogen or the loss of oxygen by an organic compound.

Reading Power

For each of the following questions, select one answer that seems most correct.

1. The main idea of section 10-10 is:

    a. Methyl alcohol is poisonous.
    b. Alcohols are hydrocarbons that contain hydroxyl (—OH) groups.
    c. The names of all alcohols end in "ol."
    d. There are many different alcohols.

2. The main idea of section 10-11 is:

    a. Carbohydrates are aldehydes and ketones.
    b. The functional group of an aldehyde is —CHO.
    c. The functional group of a ketone is —CO—.
    d. Alcohols can be oxidized to aldehydes or ketones.

3. The main idea of section 10-12 is:

    a. Carbohydrates are reduced by fermentation.
    b. Alcohols can be reduced to hydrocarbons.
    c. Yeast is needed for fermentation.
    d. Fermentation is the oldest chemical process known to civilization.

4. The main idea of section 10-13 is:

    a. All organic acids are weak.
    b. Organic acids are identified by their carboxyl group.
    c. Oxidation is the gain of oxygen or the loss of hydrogen.
    d. Ethanol oxidizes to ethanoic (acetic) acid.

5. The main idea of section 10-14 is:

    a. Esters react with water to form an acid and an alcohol.
    b. Esterification needs a strong dehydrating agent.
    c. Acids react with alcohols to form esters.
    d. Fatty acids contain carboxyl groups.

Mind Expanders

1. How many isomers can be obtained when hexane is oxidized to an alcohol? The formula for hexane is C_6H_{14}.

2. Why are all carboxyl groups found at the ends of the carbon chain (never in the middle of the chain)?

3. List the steps in the oxidation of a hydrocarbon to an acid.

4. Why is sulfuric acid needed to form esters?

5. Draw the structural formula for butyric acid C_3H_7COOH.

6. Why is industrial alcohol denatured?

7. Why do dried fruits feel sticky?

8. How do old wines change to vinegar?

Multiple-Choice

1. Which compound will react with CH_3—COOH to form the ester methyl ethanoate?

   a. CH_3—COOH   b. $CH_3CH_2CH_3$ c. CH_3OH d. CH_3COOCH_3

2. The general formula for an alcohol is:

   a. R—H   b. R—OH   c. R—COOH   d. R—COOR′

3. An example of an organic acid is:

   a. CH_3OH

   c. CH_3CO—OCH_3

   b. CH_3—CH_2—CH_2OH

   d. $C_5H_{11}COOH$

4. $CH_3CH_2COOCH_3$ is an example of an:

   a. acid   b. alcohol   c. ester   d. hydrocarbon

5. Which is an isomer of propionic acid (CH_3CH_2COOH)?

   a. $CH_2(OH)$—CO—$COOH$

   c. CH_3—CO—$CH_2(OH)$

   b. $CH_3CH_2CH_2COOH$

   d. $CH_2(OH)CH_2CH_2(OH)$

Find the Facts

Choose the sections in this chapter that best support each answer to the multiple-choice questions.

Completion Questions

Complete each of the statements on the following page using the word(s) in the list below. You may use the same word(s) more than once.

alcohol	denatured	functional group
carboxyl group	ester	organic radicals
—CHO	fermentation	reduction

1. Carbohydrates are reduced to alcohols by ⸺⸺⸺⸺ .

2. You cannot drink industrial alcohol because it is ⸺⸺⸺⸺ .

3. ⸺⸺⸺⸺ takes place when an organic compound loses oxygen.

4. R and R' represent ⸺⸺⸺⸺ .

5. ROH is the general formula for any ⸺⸺⸺⸺ .

6. An organic compound undergoes ⸺⸺⸺⸺ when it gains hydrogen.

7. A(n) ⸺⸺⸺⸺ is formed by the reaction of an acid and an alcohol.

8. The ⸺⸺⸺⸺ determines the type of organic compound it is.

9. The functional group for an organic acid is called a(n) ⸺⸺⸺⸺ .

10. The functional group for an aldehyde is ⸺⸺⸺⸺ .

True or False

Mark each statement true (*T*) or false (*F*). If the statement is false, replace the indicated word(s), using the list below, and make the statement true.

oxidation	ethanol
alcohol	carboxyl group (—COOH)

1. Pentanol is a five-carbon *alcohol.*

2. The functional group for hexanoic acid is the *hydroxyl group.*

3. We must breathe in oxygen for the continuous *reduction* of carbohydrates.

4. The *carboxyl group* is responsible for the properties of an organic acid.

5. The least poisonous alcohol is *wood alcohol.*

Table 10-14

Fill in the following table.

Type of Compound	Functional Group	General Formula
hydrocarbon	none	R—H
		R—OH
	—CHO	
		R—CO—R′
acid		
ester		

Chapter 11

What Is the Chemistry of Food?

Instructional Objectives

After completing this chapter, you will be able to:

1. Define amino acid, carbohydrate, fat, nutrient, photosynthesis, protein, and vitamin.

2. Trace all foods to green plants.

3. Recognize the sun as our energy source.

4. List six nutrients and explain the role of each (including each vitamin) in sustaining life.

5. Identify food sources for each nutrient and vitamin.

6. Compare different methods of cooking food with respect to nutritional and chemical changes.

7. List five ways to preserve food.

Chapter 11 Contents

	Introduction	267
11-1	Where Does Food Come From?	267
11-2a	What Do We Need from Food?	268
11-2b	How Are Carbohydrates Essential for Life?	268
11-2c	How Are Proteins Essential for Life?	270
11-2d	How Are Fats Essential for Life?	271
11-2e	How Are Minerals Essential for Life?	272
11-2f	How Were Vitamins Discovered?	272
11-2g	How Is Vitamin A Essential for Life?	273
11-2h	How Are the B-Complex Vitamins Essential for Life?	273
11-2i	How Is Vitamin C Essential for Life?	274
11-2j	How Is Vitamin D Essential for Life?	274
11-2k	How Is Vitamin E Essential for Life?	274
11-2l	How Is Vitamin K Essential for Life?	275
11-3a	How Does Cooking Change Our Food?	275
11-3b	What Happens to Food When We Fry It?	275
11-3c	What Happens to Food in Baking?	276
11-3d	How Does Boiling Change Food?	278
11-3e	How Do Microwave Ovens Cook Our Food?	278
11-4	How Can Food Be Preserved?	279
	Written Exercises	282
	Career Information	288

Chapter 11
What Is the Chemistry of Food?

Introduction

The human body's need for energy and growth is met by the foods we eat. Food is a source of life and wealth. Many people in the world die each day from starvation, and the situation is expected to worsen. In the wealthy countries of the world, there is an overabundance of food, but many people do not know how to eat properly. There are four major food-exporting countries in the world today: the United States, Canada, Australia, and Argentina.

11-1   Where Does Food Come From?

Every food can be traced back to a green plant. They produce complex molecules from simple compounds and energy.

Table 11-1

$$n\,CO_2 \;+\; n\,H_2O \;\xrightarrow{\quad\text{SUNLIGHT}\quad}\; Cn(H_2O)_n \;+\; n\,O_2$$

carbon   water   carbohydrate   oxygen

Green plants are our food producers.

The process by which green plants make food is called **photosynthesis**. *Photo* is the Greek word for "light," and *synthesis* means "to put together." Photosynthesis is the uniting of carbon dioxide and water with the help of sunlight, to produce carbohydrates.

The chemicals needed to make food come from the earth itself. If living things never died, the resources of the earth would become exhausted. Death is nature's way of recycling the resources that life has borrowed from the earth. Our food comes from other living things. The essential element present in all living things is carbon. This element must be consumed in the foods we eat. Carbon is not an abundant element on this planet (see Chapter 10, Introduction), or in the universe. The supply of carbon must be renewed to the earth. Living things must die to ensure the continuation of life in the future.

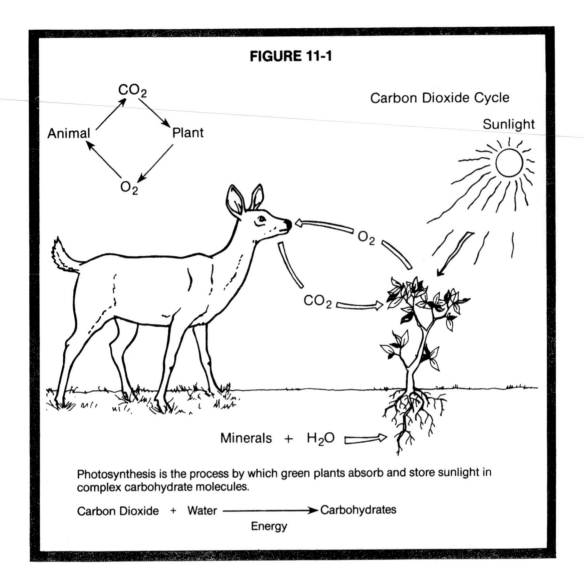

FIGURE 11-1

Carbon Dioxide Cycle

Photosynthesis is the process by which green plants absorb and store sunlight in complex carbohydrate molecules.

Carbon Dioxide + Water ⟶ Carbohydrates
Energy

11-2a   What Do We Need from Food?

Our body requires six major components from food to sustain life. These components, called **nutrients**, are **carbohydrates**, **proteins**, **fats**, **minerals**, **water**, and **vitamins**.

11-2b   How Are Carbohydrates Essential for Life?

The word carbohydrate has two parts. *Carbo* means "carbon" and *hydrate* means "water." A carbohydrate is a compound in which carbon is combined with

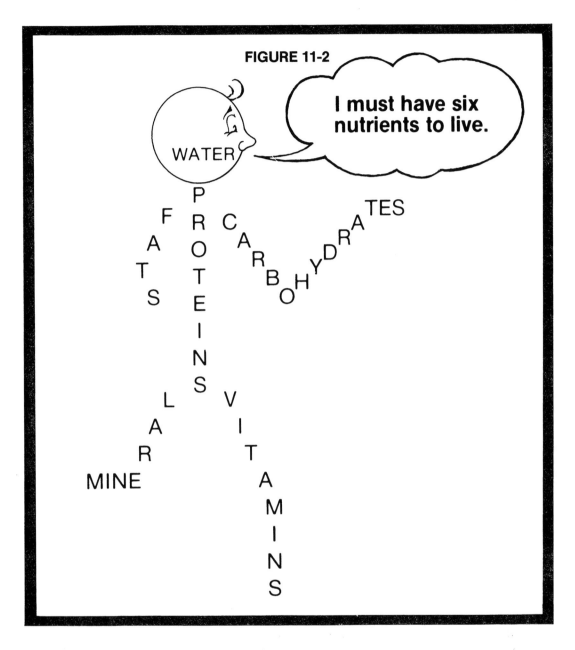

FIGURE 11-2

I must have six nutrients to live.

hydrogen and oxygen in the same proportion as in water. The general formula for a carbohydrate is $C_n(H_2O)_n$. They take different forms: wood (or cellulose), sugar, and starch. Some living things (termites) are able to digest wood. Their intestines contain the bacteria that digest cellulose. These bacteria cannot survive in human intestines. Through digestion, all carbohydrates are reduced to glucose, which is the sugar found in our blood. Glucose is burned in our cells to release energy. Starch and sugar are essential in human nutrition, and are the source of about 80% of our body's energy. Starvation, or diets that deprive the body of carbohyrates (such as the "high protein" diet), can cause serious harm. When our bodies are deprived of needed carbohy-drates, we must substitute fat as our main energy source. Our body cells burn car-bohydrates (glucose) very efficiently, but fats burn more slowly. The result may be a disease called **acidosis**. Diabetes is a disease in which glucose is not able to be transferred from the blood to the body cells. Diabetics may suffer from acidosis as a complication of their disease. Refined sugars, such as sucrose (cane sugar sold in supermarkets) are poor sources of human nutrition. Refined carbohydrates cause the release of cholesterol into the blood. Cholesterol is a major contributor to heart disease. Complex carbohydrates in the form of potatoes, pasta, rice, bread, and whole-grain cereals are a wholesome and essential part of human nutrition.

Table 11-2

$$n\ H_2O\ +\ (C_6H_{10}O_5)_n \overset{\text{DIGESTION}}{\longrightarrow} n\ C_6H_{12}O_6$$

starch   glucose

When food is digested its nutrients are reacted with water.

Table 11-3

$$C_6H_{12}O_6\ +\ 6\ O_2 \longrightarrow 6\ H_2O\ +\ 6\ CO_2\ +\ \text{energy}$$

Glucose fuels our cells.

11-2c   How Are Proteins Essential for Life?

All living things are composed of proteins. We must consume proteins to provide the chemicals needed to grow and repair body tissue. When you are past your growth years, you still lose dead skin, hair, blood, and other tissue. Proteins are needed to replace lost tissues. Just as your home may be constructed of bricks, the proteins in your body are constructed of small molecular units called **amino acids**. Just as every person's fingerprints are unique, every person possesses proteins which are different from everyone else's. No two persons can have the same proteins, except for identical twins. Every animal makes proteins from amino acids. The proteins you eat are broken down to amino acids through digestion. They are then absorbed into your blood and carried to your cells. These amino acids are then recombined to produce the special proteins needed by your body. Proteins can be obtained only from other living things.

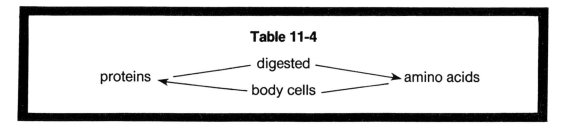

Table 11-4

proteins ⟷ digested / body cells → amino acids

Animal protein is derived from fish, fowl, meat, eggs, and milk products. It is complete protein because it provides all the essential amino acids needed to make human protein. **Nonessential amino acids** can be made by our own cells. Vegetable or plant proteins are incomplete because they have some, but not all, of the essential amino acids; in order to get complete protein from vegetables, you must combine them, such as beans and rice, etc. **Essential amino acids** are those amino acids needed by our body, but which cannot be made by our cells. An expensive steak has the same quality protein as cottage cheese. They both contain animal protein. A starvation diet results in the body burning its own protein for energy. This process consumes lean body tissue (muscles and organs), which can cause irreversible damage to vital organs, (heart, liver, kidneys, and brain).

Foreign proteins (proteins not made by your own cells) are rejected by your body. This is a major problem faced in organ transplants. Animals and insects that carry protein in their saliva (mosquitoes, for example) may inflict a poisonous bite or cause an allergic reaction. The mosquito, in its zeal to extract blood, injects its saliva into its victim.

11-2d   How Are Fats Essential for Life?

Fats are composed of long chains of carbon and hydrogen atoms. Fats that have only single bonds between carbon atoms are **saturated fats**. Those that have many double bonds between carbon atoms are called **polyunsaturated fats**. Oil is a liquid fat. Saturated fats may interfere with your body's ability to cope with cholesterol, which can result, over a period of time, in heart disease. The polyunsaturated fats, on the other hand, have been found to reduce cholesterol levels in the blood. Hydrogenated vegetable oils are polyunsaturated oils that have hydrogen added to their double bonds.* Hydrogenated vegetable oils can be turned into saturated fats (see section 10-7).

Polyunsaturated oils such as corn oil and safflower oil are vulnerable to oxidation near the double bond. They oxidize by combining with atmospheric oxygen. This element is second only to fluorine in its strong attraction for electrons. This electron-hungry nonmetal seeks out the electron-rich double bond and adds onto the carbon chain. When this happens, the oil turns rancid, and food prepared with it tastes spoiled. Rancidity can be avoided by packing the oil in air- and light-tight containers. Antioxidants such as BHA, BHT, and vitamin E, and spices such as sage and rosemary, are often added to foods to retard oxidation.

*Saturated oils (coconut, palm, and hydrogenated vegetable oils) are often used in baking breads and cakes. These baked products have a longer shelf life in supermarkets because saturated oils resist rancidity. Products baked with polyunsaturated oils are more healthful.

Fats are burned for energy in your body. Fat also insulates and protects your tissues. Many other food components, such as lecithin and vitamins A, D, K, and E are soluble in fats. These components can only be obtained from the oil portion of foods. Fats are digested to fatty acids and glycerol.

Table 11-5

| FAT | + | WATER | DIGESTIVE ENZYME | → | FATTY ACID | + | GLYCEROL |

11-2e   How Are Minerals and Water Essential for Life?

Calcium and phosphorus are needed for strong bones and teeth. Apatite (calcium phosphate) is a mineral found in some stones. When placed into broken bones, it is found to stimulate bone growth and is absorbed by the bone tissue. Milk is a good source of calcium. All tissues require water and minerals. Blood cannot clot without calcium. Iron and trace amounts of copper are needed by red blood cells. Iron is found in liver and yeast. You may survive for three to five weeks without food, but you cannot live without water for more than a few days.

11-2f   How Were Vitamins Discovered?

It had been known for many years that disease can result from food deficiency. Sailors discovered that scurvy could be prevented by eating fresh fruits and vegetables on long ocean voyages. In 1882, a Japanese naval doctor, Takaki, discovered that beriberi could be prevented if the sailors were required to eat meat, barley, and fruits. In 1897, a Dutch doctor, Christiaan Eijkman, discovered that beriberi could be cured if the brown polishings (skins) of rice were eaten. In 1911, the American scientist Casimir Funk proposed his theory of vitamins in food. He said that foods contain vitamins that prevent certain diseases. In England in 1912, an important experiment proved the existence of vitamins. Rats were fed a purified diet of fat (lard), carbohydrate (starch), protein (casein or milk protein), water, and minerals. The diet proved to be deficient and the rats became ill. On the addition of small amounts of whole milk, the rats recovered. At this point (1915), the hunt for vitamins was on.

11-2g   How Is Vitamin A Essential for Life?

Vitamin A was the first fat-soluble food factor identified. The second was water-soluble B, or **vitamin B**. Lack of vitamin A results in a loss of weight, and an increased chance of infection. Vitamin A prevents the eye disease called xerophthalmia, in which the eye tissues dry up. It also prevents night blindness, which is a loss of vision at twilight and dawn. This vitamin helps your eyes adjust to the dark more rapidly. It is stored in your liver. You should be aware that too much vitamin A can be poisonous. The English health-food enthusiast Basil Brown actually overdosed on a combination of vitamin A and carotene. In February 1974, Basil consumed more than 10,000 times the recommended amount of this vitamin. Within 10 days, his skin turned yellow and he died. Shark and polar bear liver are poisonous because of the large concentration of vitamin A found in them. Recent findings indicate that retinoic acid, derived from vitamin A, rubbed into acne, produces very good results. This vitamin interferes with **vitamin K** to a small extent. In a normal diet, this interference is hardly noticeable. Vitamin A can be found in liver, egg yolk, milk (cream), and butter. Yellow vegetables (carrots) contain **carotene**. Carotene is vegetable vitamin A, and must be digested to be changed into vitamin A for use in human nutrition.

Table 11-6

$$C_{40}H_{56} + 2H_2O \longrightarrow 2C_{20}H_{29}OH$$

carotene   vitamin A

In general, the more sunlight a plant receives, the more pigment (**chlorophyll** and **carotenoids**) will be found in their leaves and stems. This pigmentation enables them to cope with the large input of solar energy they receive. Dark green and orange vegetables will be richer in carotene. Unfortunately, the darker leaves are generally unattractive because they are on the outside of the vegetable and exposed to insects and other abuses.

11-2h   How Are the B-Complex Vitamins Essential for Life?

At first, vitamin B was thought to be only one vitamin. It was soon discovered that there are many B vitamins. Vitamin B-1, or **thiamine**, prevents the disease beriberi. This disease causes paralysis, anemia, and weakness. If it is not treated, death will result. Brewer's yeast and wheatgerm are rich in B-complex vitamins. Milk, whole-grain cereals, and brown rice are also good sources for B-complex vitamins. Vitamin B-1 is sensitive to heat, and about 20–50% of the vitamin is lost in cooking. Some vegetables, such as soybeans, red cabbage, beets, brussels sprouts, and some berries, carry proteins that will neutralize vitamin B-1. No cases of vitamin B-1 deficiency have been reported due to these vegetables. The B-complex vitamins are all water soluble. Riboflavin is sensitive to alkali (bases). Cooking with sodium bicarbonate ($NaHCO_3$) will make some foods appear bright and colorful. They make carrots a bright orange, and green peas a bright green, but the bicarbonate will destroy the riboflavin. This vitamin is important in preventing skin disorders. **Niacin**, another B vitamin, prevents pellagra. Pellagra means "rough skin," in Italian. This is a disease that afflicted many poor people in the southern part of the United States. Dr. Joseph Goldberger, of the U.S. Public Health Service, found that pellagra could be prevented with niacin.

11-2i How Is Vitamin C Essential for Life?

In 1772, Dr. James Lind, an English surgeon, discovered that scurvy could be prevented if sailors were required to eat citrus fruits. Scurvy is a disease in which there is bleeding from the gums and under the skin. It also causes teeth and gums to decay. If untreated, death will result. To this day, English sailors are nicknamed "limeys" because they were required to drink lime juice. Dr. Linus Pauling, an American Nobel Prize chemist, and many other people, believe that vitamin C will prevent the common cold. This belief has not been proven true.

Vitamin C, or ascorbic acid, is contained in all citrus fruits, tomatoes, strawberries, cabbage, melons, and green peppers. It is water soluble. It cannot be stored in the body. It must be consumed fresh every day. On the day that you are deprived of this vitamin, you are more vulnerable to infection and dental caries (cavities). Vitamin C will be destroyed by heat and exposure to air. Orange juice that stands for a long time will lose some of its vitamin C content. Green plants make vitamin C from sugar. As a rule, the more sunlight the plant receives, the darker its leaves, and the greater its vitamin C content.

11-2j How Is Vitamin D Essential for Life?

Vitamin D is sometimes called the "sunshine vitamin." Ergosterol and cholesterol, found under the skin, are changed into vitamin D on exposure to the ultraviolet rays of the sun. Vitamin D is fat soluble. If taken in large doses, vitamin D can be poisonous. This vitamin is important for strong bones and teeth. Vitamin D prevents the disease called rickets. This is a disease in which the bones soften and are easily bent out of shape. Vitamin D is found in milk (cream) and in all fish oils.

11-2k How Is Vitamin E Essential for Life?

We know that vitamin E is needed by rats for reproduction. The need for vitamin E in human nutrition has never been established. Vitamin E is fat soluble, and is found in green lettuce leaves, whole-grain cereals, beef liver, egg yolk, and wheatgerm oil. It may be that plants manufacture this vitamin to prevent the oil in their seeds from turning rancid. Vitamin E is an antioxidant for polyunsaturated oils.

11-21 How Is Vitamin K Essential for Life?

Vitamin K is a fat-soluble vitamin that is involved with the clotting of blood. Three materials must be present for blood to clot: platelets (cells found in blood), the mineral calcium, and vitamin K. The letter "K" comes from the German word *Koagulazion*, which means "clot." Vitamin K is found in green leafy vegetables such as spinach, alfalfa, cabbage, kale, and cauliflower. Very little is found in tomatoes and fruits.

11-3a How Does Cooking Change Our Food?

There are several ways to prepare foods. We can fry, broil, bake, roast, or boil foods. Cooking changes the appearance and flavor of foods. It also changes the nutritional value of foods. Heat will coagulate proteins. This will make the protein easier to digest. Cooked eggs have several advantages over raw eggs. Of course raw eggs are not palatable for most people. Uncooked eggs contain the protein avidin, which destroys **biotin** (a B-complex vitamin). Eggs also have ovomucoid, a protein which inhibits **trypsin**. Trypsin is an enzyme needed in the digestion of proteins. Both of these proteins are denatured (coagulated) by cooking. Proteins are normally coagulated by the hydrochloric acid in our stomach. Consuming coagulated protein eases the digestive process. We will consider the effects of frying, baking, and boiling foods.

11-3b What Happens to Food When We Fry It?

When food is fried, it is heated in oil or fat. The proteins are coagulated by the heat. The coagulation, or clumping, of proteins causes them to become insoluble in water. Frying (and hard boiling) an egg is an example of protein coagulation. The fats and oils in which foods are fried also undergo chemical changes. The fats are absorbed by the food being fried. This absorption makes the fried food more difficult to digest. Proteins begin to be digested in the stomach, and that digestion is completed in the small intestine. Fats begin to be digested in the small intestine. The fats absorbed by the foods inhibit the start of protein digestion. Many vitamins (B-complex and C) are destroyed by the heat of frying. If fried foods are overheated, the oils break down and free fatty acids are released into the food. Irritating acrolein is formed from glycerol, and the food becomes distasteful. Acrolein is responsible for odors and eye irritation that often accompanies fried foods.

Table 11-7

$$
\begin{array}{c}
\text{H} \\
| \\
\text{H—C—OH} \\
| \\
\text{H—C—OH} \\
| \\
\text{H—C—OH} \\
| \\
\text{H}
\end{array}
\quad\text{HEAT due to FRYING}\quad\longrightarrow
\begin{array}{c}
\text{H} \quad\quad \text{H} \\
\diagdown \quad\quad \diagup \\
\text{C}=\text{C—C} \quad + \quad 2\,H_2O \\
\diagup \quad | \quad \diagdown \\
\text{H} \quad \text{H} \quad \text{O}
\end{array}
$$

GLYCEROL   ACROLEIN
(from oil)

11-3c   What Happens to Food in Baking?

Baking takes place in a hot oven between 250°F and 450°F. This method of preparing food is inefficient. The air in the oven is so thin at high temperatures that few hot air molecules come in contact with the food. In comparison, boiling food is touched on all sides by dense hot water. This is why boiled potatoes cook faster than baked potatoes, and why breads, cakes, and pies require high temperatures for long periods of time to bake.

Bread is a basic baked food. Bread is a dough made from a grain flour and water. It has a leavening agent added to it which makes the dough rise. Bread can be leavened with yeast or with baking powder. Each method produces carbon dioxide gas (CO_2), which causes the bread to rise, and makes the bread easier to chew and digest.

When yeast is used, salt is sometimes added to regulate the rate of fermentation. Salt slows fermentation, so that the bread will not be too porous or "light." The yeast consists of microscopic plant cells that ferment the starch and added sugar into alcohol and carbon dioxide. As the temperature inside the dough reaches 175°F, the yeast cells are killed. Fermentation ends, but the alcohol vaporizes and aids the leavening. When the temperature reaches 212°F, the water vapor also helps to leaven the bread.

Table 11-8

$$C_6H_{12}O_6 \xrightarrow[\text{heat}]{\text{yeast}} 2\,C_2H_5OH \;+\; 2\,CO_2\uparrow$$

sugor $\xrightarrow{\text{FERMENTATION}}$ alcohol   +   carbon
(or starch)   dioxide

Steps in the Commercial
Baking of Bread

FIGURE 11-3a
Dough making

Figure 11-3b
Refrigeration to retard fermentation

FIGURE 11-3c
Moist heat to accelerate fermentation

FIGURE 11-3d
Baking

Courtesy:
Orza Bakers
Yonkers, NY

The chemistry of baking is similar to that of winemaking. Yeast ferments sugars and starches into ethyl alcohol and carbon dioxide. In winemaking, the alcohol is the major product. In baking, the carbon dioxide is of prime importance.

Baking powder is a mixture of a weak acid and a weak base. When these powders are mixed with water in the preparation of dough, carbon dioxide is produced. The gas causes the bread to rise.

The heat of baking partly changes starches into sugar. Baked sweet potatoes are sweeter than raw yams for this reason.

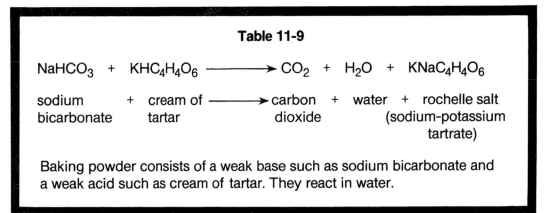

Table 11-9

$$NaHCO_3 + KHC_4H_4O_6 \longrightarrow CO_2 + H_2O + KNaC_4H_4O_6$$

sodium + cream of ⟶ carbon + water + rochelle salt
bicarbonate   tartar   dioxide   (sodium-potassium
                                                          tartrate)

Baking powder consists of a weak base such as sodium bicarbonate and a weak acid such as cream of tartar. They react in water.

11-3d   How Does Boiling Change Food?

Vegetables are composed of cells surrounded by wood or cellulose walls. These cells are "glued" to each other by a jelly-like carbohydrate called **pectin**. When vegetables are boiled in water, the pectin dissolves and the plant loses its rigid structure. Pectin is quickly lost from vegetables if heated in an alkaline solution. On the other hand, pectin becomes a thick gel in an acidic medium. This is the reason that pectin is added to jam and jelly. When vegetables ripen or are boiled, they lose the pectin between their cell walls and become soft.

11-3e   How Do Microwave Ovens Cook Our Food?

Microwaves are a form of radiant energy and are invisible to the human eye. They are often used for long-distance telephone communication, and for cooking foods.

Microwave energy is useful in cooking because it is absorbed by water and converted into heat inside food. The microwave oven contains a magnetron tube and is lined with metal inside its walls. The magnetron tube sends out (transmits) the microwave energy. This energy stays inside the oven because it cannot penetrate the metal walls. The microwaves are reflected back into the oven, where the food is placed.

Water in the food absorbs the microwaves and becomes hot. The water boils off as steam. The result is that the food is quickly cooked internally as if it were steamed. Since the air inside the oven remains cool, the food is not baked.

The only danger of this form of cooking is the leakage of microwaves into the area outside the oven. If this happens, the microwaves could be absorbed by persons who are exposed to the waves. Though the power level is too low to cause immediate injury, persons with heart pacemakers must be very cautious near microwave ovens.

Microwave ovens are generally safe. Old ovens should be kept clean, especially around the door, and checked for leakage. Metals must never be placed inside the oven. They would reflect the microwaves back into the magnetron tube and cause it to "burn out."

Follow these steps to avoid exposure to microwaves:

1. Microwave ovens automatically turn off when the door is open. Do not attempt to operate the oven with the door open.

2. Do not place any objects between the door and the front face of the oven. Clean away dirt and cleanser residue from the sealing surfaces. Door seals and sealing surfaces should be clean and in good condition.

3. Do not operate the oven if it is damaged. The door should not be bent. The hinges and latches should not be broken.

11-4   How Can Food Be Preserved?

Microorganisms need food sources to sustain their lives. Meat and vegetables will rot or decay due to the action of microorganisms. Foods will be preserved if the action of these bacteria and protozoa are stopped or slowed down. This task can be accomplished by drying the food (dried fruits and dried meats). One way of drying meat is by salting it. The salt will draw the moisture from the meat. Microorganisms must have water to consume or decay the food. Cooking will help preserve food because heat destroys microorganisms. Refrigeration slows down microorganic activity. Strong exposure to nuclear radiation is also used to kill microorganisms. In this way, foods can be safely preserved without refrigeration. Canned foods are cooked in their cans so they can stand for long periods of time without spoiling. Swollen cans may indicate that all the microorganisms in the food were not killed. A serious disease such as **botulism** can result from the consumption of the spoiled food from swollen cans.

Now You Know

1. The world is suffering from a lack of food. The shortage is expected to get worse in the future.

2. The United States, Canada, Australia, and Argentina are the four major food-exporting countries in the world.

3. The source of all food is the green plant. All foods can be traced back to the green plant.

4. Green plants make their food from simple compounds and sunlight by the process of photosynthesis.

5. We need food in order to grow, repair, and replace tissues. Food is our energy source.

6. We must have six nutrients from food. These nutrients are proteins, fats, carbohydrates, vitamins, water, and minerals.

7. Carbohydrates consist of sugar and starch. They are our body's chief energy source.

8. Carbohydrates are broken down into glucose through digestion and burned in the cells of our body.

9. Proteins are found in all living things.

10. Proteins are broken down into amino acids through digestion.

11. Amino acids are used in your cells to construct your body's own protein.

12. No two people, except identical twins, can have the same body proteins.

13. Essential amino acids are those our body cannot make for itself. They are needed for protein synthesis.

14. Fats are important because they provide energy, tissue protection, and essential food components, such as vitamins A, D, K, and E.

15. Minerals are important for bones, teeth, blood, and other tissues.

16. Vitamins are needed in small amounts in our diet. We cannot live without vitamins.

17. Vitamin C is a water-soluble vitamin that is destroyed by heat, air, and alkali. It cannot be stored in the body. It must be consumed fresh every day. It helps your body to resist infection and heal wounds.

18. The vitamin B-complex consists of many water-soluble vitamins. They prevent several diseases, including beriberi and pellagra. B-complex vitamins are found in brewer's yeast, wheatgerm, whole-grain cereals, and milk.

19. Vitamin A is needed for your skin and eyes. It is found in the oil part of certain foods such as fish-liver oils, liver, and milk.

20. Vitamin D is a fat-soluble vitamin that is important for strong bones and teeth. It is found in fish oils. It is sometimes called the "sunshine vitamin" because sunlight changes chemicals in the skin into the vitamin.

21. Vitamin E has not been proven to be essential to human nutrition. It is fat soluble and found in egg yolk, green lettuce leaves, and wheatgerm oil.

22. Vitamin K is one of the three essential materials needed for the clotting of blood. It is found in spinach, cabbage, and kale.

23. Fried foods are more difficult to digest because the fats in which they are fried are absorbed by the food and inhibit protein digestion in the stomach.

24. The protein in fried foods is coagulated. This is not harmful because coagulation is the first step in normal protein digestion.

25. When oil is overheated in the frying process, it breaks down into acrolein. Acrolein is an irritant.

26. When bread is baked, a dough made of grain flour is mixed with a leavening agent and heated in an oven.

27. Water in food absorbs microwaves and becomes hot, so the food cooks quickly internally.

28. Foods are preserved by killing or controlling the activity of microorganisms. The methods used are refrigeration, drying, salting, radiation, and cooking.

New Words

amino acids	The molecular units that make up proteins.
carbohydrate	Sugar or starch. The chief source of energy for our body.
fats	An energy-rich substance found in plants and animals. They are composed of fatty acids combined with glycerol.
nutrient	A food component needed to sustain life.
photosynthesis	The process by which green plants make food from simple compounds, with the help of sunlight.
protein	The nutrient needed for the growth and repair of tissues. Proteins are made of small units called amino acids. All living things are made of proteins.
vitamin	A nutrient needed in small quantities to sustain life.

Reading Power

For each of the following questions, select one answer that seems most correct.

1. The main idea of the introduction is:

   a. Food is needed for energy and growth.
   b. There is a world food shortage that is expected to get worse in the future.
   c. The United States is a major food-exporting nation.
   d. All of the above.

2. The main idea of section 11-1 is:

   a. Food is needed for energy and growth.
   b. Photosynthesis is the union of carbon dioxide and water.
   c. The source of our food is the green plant.
   d. Food comes from sunlight.

3. The main idea of section 11-2a is:

   a. We need six nutrients from food.
   b. We need carbohydrates from food.
   c. We need fats from food.
   d. We need proteins from food.

4. The main idea of section 11-2b is:

   a. Carbohydrates are compounds of carbon, hydrogen, and oxygen.
   b. Carbohydrates consist of sugars and starches.
   c. The chief energy source for your body is carbohydrates.
   d. Fats can be substituted for carbohydrates.

5. The main idea of section 11-2c is:

   a. Proteins are needed to build and repair body tissue.
   b. Proteins are synthesized in the cells of your body.
   c. Animal proteins are needed because they provide all essential amino acids.
   d. Vegetable proteins do not provide all essential amino acids.

6. The main idea of section 11-2d is:

   a. Vegetable proteins do not provide all the essential amino acids.
   b. Polyunsaturated fats can reduce blood cholesterol.
   c. Hydrogenated vegetable oils have some double bonds changed into single bonds.
   d. Fats are composed of energy-rich molecules that may contain vital food components.

7. The main idea of section 11-2e is:

   a. Water is the most essential food component in your diet.
   b. Minerals are needed for strong bones and teeth.
   c. Calcium is the most difficult mineral to obtain for our body.
   d. Iron is needed by our red blood cells.

8. The main idea of section 11-2f is:

   a. Vitamins are vital for life.
   b. Vitamins prevent diseases.
   c. Vitamins are needed in small amounts.
   d. Vitamin pills should be taken every day.

9. The main idea of section 11-2g is:

   a. Vitamin A is a fat-soluble vitamin.
   b. Vitamin A was the first vitamin isolated.
   c. Vitamin A is essential to life, but an overdose is poisonous.
   d. Vegetable vitamin A, or carotene, must be digested into vitamin A.

10. The main idea of section 11-2h is:

    a. There are many B-complex vitamins.
    b. Each B-complex vitamin is needed to sustain life.
    c. The richest source of B-complex vitamins is brewer's yeast.
    d. B-complex vitamins are destroyed by heat.

11. The main idea of section 11-2i is:

    a. The effects of vitamin C were discovered in 1772.
    b. Vitamin C is needed to sustain human life.
    c. Vitamin C is needed fresh every day.
    d. It has not been proven that vitamin C prevents colds.

12. The main idea of section 11-2j is:

    a. Vitamin D is needed for the proper utilization of calcium and phosphorous in the body.
    b. Vitamin D is a fat-soluble vitamin.
    c. Vitamin D is found in fish oils and butterfat.
    d. Vitamin D is made in your body on exposure to sunlight.

13. The main idea of section 11-2k is:

    a. Vitamin E is a fat-soluble vitamin.
    b. Vitamin E is an antioxidant used in polyunsaturated fats.
    c. Rats need vitamin E.
    d. Vitamin E is found in beef liver.

14. The main idea of section 11-2l is:

    a. Vitamin K is needed to clot blood.
    b. Three factors are needed to clot blood.
    c. Vitamin K is a fat-soluble vitamin.
    d. Coagulation means "clot."

15. The main idea of section 11-3a is:
    a. Cooking causes changes in food.
    b. Cooking can make digestion of food easier.
    c. There are many ways to cook food.
    d. All of the above.

16. The main idea of section 11-3b is:
    a. Fried foods are harder to digest.
    b. Acrolein is a poisonous substance that comes from the breakdown of fats in the frying process.
    c. Many vitamins are destroyed when foods are fried.
    d. Proteins are coagulated, but they are more difficult to digest in fried foods.

17. The main idea of section 11-3c is:
    a. Baking is a more healthful way to prepare foods.
    b. Baking takes place in an oven.
    c. Baking causes fermentation in dough and the conversion of starch into sugar.
    d. Yeast consists of microscopic plants that ferment carbohydrates.

18. The main idea of section 11-3d is:
    a. Plant cells are "glued" together with pectin.
    b. Vegetables are softened when boiled in water because they lose their pectin.
    c. Pectin is soluble in an alkaline solution.
    d. Rotting fruits and vegetables grow soft.

19. The main idea of section 11-3e is:
    a. Microwaves may escape from a microwave oven.
    b. Microwave energy cooks food quickly because it is absorbed by water and changed into heat.
    c. Microwaves are a form of radiant energy.
    d. Microwave ovens should be kept clean.

20. The main idea of section 11-4 is:
    a. Some ways of preserving foods are by refrigeration, radiation, cooking, and drying.
    b. When foods rot or spoil it is because of the actions of microorganisms.
    c. Foods are preserved by the cooking process.
    d. All of the above.

Mind Expanders

1. Why is baked food more healthful than fried food?

2. What must we derive from food to maintain good health?

3. How is each nutrient essential to your health?

4. How do we derive energy from food?

5. How are proteins synthesized in our cells?

6. Why must we have essential amino acids in our diet?

7. Describe four ways of preserving food.

8. Why did the Indians hang up their meat and fish to dry?

9. How does smoking foods preserve them?

Complete the Following Table

Vitamin	Importance	Food Sources	Miscellaneous Information
	prevents scurvy		
	prevents night blindness		
B-1 (thiamine)			
D			
E			
K			
niacin			
riboflavin			

Completion Questions

Complete each of the following statements, using the word(s) below. The same word(s) may be used more than once.

carbohydrates	vitamin A	vitamin E
green plants	vitamin C	essential amino acids
vitamin K	protein	amino acids

1. All food on the earth can be traced to the _____ .

2. The nutrient needed for tissue growth and repair is _____ .

3. _____ is the chief energy source for living things.

4. Proteins in our bodies are constructed from _____ .

5. _____ cannot be made by your body.

6. _____ cannot be stored in your body and must be consumed fresh each day.

7. _____ has not been proven to be needed in human nutrition.

8. _____ when taken to excess is poisonous.

9. _____ is needed to clot blood.

10. The oils in fried foods hold back the digestion of _____ .

Multiple-Choice

1. All of the following are food-exporting countries except:
   a. the United States
   b. Canada
   c. the Soviet Union
   d. Australia

2. Our food supply depends on all of the following except:
   a. green plants   b. sunlight   c. minerals   d. animals

3. All of the following are nutrients except:
   a. protein   b. vegetables   c. fats   d. minerals

4. All of the following are carbohydrates except:

    a. cellulose   b. meat   c. starch   d. sugar

5. All of the following are animal protein sources except:

    a. soybeans   b. cottage cheese   c. steak   d. skim milk

6. In digestion, all carbohydrates are broken down into:

    a. amino acids   b. glucose   c. glycerol   d. fatty acids

7. Leavening agents cause bread and cakes to rise by the formation of:

    a. carbon dioxide   b. baking powder
    c. sour milk   d. yeast

8. Foods that are the most difficult to digest are prepared by:

    a. broiling   b. frying   c. baking   d. steaming

9. Foods can be preserved in all of the following ways except:

    a. soaking in warm water   b. refrigeration
    c. cooking   d. salting

True or False

Mark each of the following statements true (*T*) or false (*F*). If the statement is false, replace the indicated word(s), using the list below, to make the statement true.

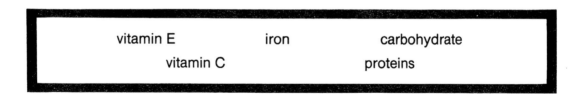

vitamin E	iron	carbohydrate
vitamin C		proteins

1. $C_n(H_2O)_n$ is the general formula for a *fat*.

2. Vitamin C is found in oranges and prevents *scurvy*.

3. *Vitamin B* has not been proven to be essential to human life.

4. Cooked foods are easier to digest because the *fats* are coagulated by heat.

Find the Facts

Choose the section in Chapter 11 that supports each answer to your true or false questions. Write the section number next to each answer.

Matching

Write the number in column B that relates to the word(s) in column A in the spaces indicated.

COLUMN A	COLUMN B
_____ nutrients	1. An energy-rich material found in plants and animals.
_____ amino acids	2. The process by which green plants make food.
_____ fats	3. The food components needed for life.
_____ photosynthesis	4. A fat-soluble vitamin that prevents soft bones.
_____ proteins	5. The molecular units that make up proteins.
_____ vitamin C	6. A nutrient needed to build and repair tissues.
_____ vitamin D	7. A substance that prevents scurvy.

Science Careers in the Food Industry

High School Diploma	2-Year Degree	4-Year or Graduate Degree
cook	chef	food chemist
baker	pastry chef	
	dietitian assistant	dietitian
	lab technician	

Chapter 12

How Are Chemical Creations Changing Life-Styles?

Instructional Objectives

After completing this chapter, you will be able to:

1. Define addition polymerization, amino acid, cold-setting plastic, condensation polymerization, copolymer, elastomer, macromolecule, monomer, plastic, silicone, thermoplastic polymer, thermosetting polymer, vulcanization.

2. Compare different kinds of polymers (thermosetting, thermoplastic, elastomer, and silicone).

3. Compare natural and synthetic polymers.

4. Recognize addition and condensation polymerization reactions.

5. Discuss problems relative to the disposal of polymers.

Chapter 12 Contents

	Introduction	291	
12-1	What Are Plastics?	291	
12-2a	What Kinds of Polymers Are There?	292	
12-2b	What Are Thermoplastic Polymers?	292	
12-2c	What Are Thermosetting Polymers?	292	
12-2d	What Are Elastomers?	293	
12-2e	What Are Fibers?	293	
12-3a	How Are Polymers Formed?	297	
12-3b	How Are Polymers Made in Addition Reactions?	298	
12-3c	How Is Synthetic Rubber Made?	300	
12-3d	How Are Polymers Made by Condensation?	300	
12-4	What Are Silicones?	301	
12-5	How Can Synthetic Polymers Be Disposed Of?	302	
	Written Exercises	304	

Chapter 12

How Are Chemical Creations Changing Life-Styles?

Introduction

In Chapter 4, we learned that atoms combine to form molecules. In this chapter, we will learn how chemists combine molecules to create new substances.

After the American Civil War, an award of $10,000 was offered to any person who could find a substitute for ivory. This material was used in the manufacture of billiard balls. Ivory had to be imported from Africa, and it was expensive. John Wesley Hyatt of Albany, New York, set out to accomplish this task. He combined cellulose (cotton) with nitric acid, camphor, and alcohol at high pressure and made a clear plastic material. He called it **celluloid**. This was the first plastic made by a person. It was not a suitable substitute for ivory. It shrank in its mold, and was difficult to machine. Hyatt did not collect the award money, but he opened up the age of plastics. He went into business making dental plates, shirt collars, cuffs, combs, and toys from celluloid. It opened up the movie industry because celluloid was used to make movie film. Celluloid is very flammable, however, and there was constant danger of fire in the movie theaters. In the 1920s, a cellulose acetate film was developed that was more resistant to burning. In this chapter, we will see how the scientist, in his quest to make new chemical creations, copies from and often improves on natural products.

12-1   What Are Plastics?

Many natural materials such as wood, cotton, hair, wool, silk, and rubber are composed of giant molecules that have molecular weights of 1,000,000 atomic mass units. These **macromolecules** may contain 200,000 atoms. They are composed of small molecular units that are linked together in almost endless chains. Each unit is called a **monomer**. *Mono* means "one," *mer* is the Greek word for "unit," and *poly* means "many." Many monomers combine to make a **polymer**. The plastics that we have today are all polymers. Besides the natural polymers, we have synthetic polymers such as nylon, polyester, polyethylene, polystyrene, and synthetic rubber.

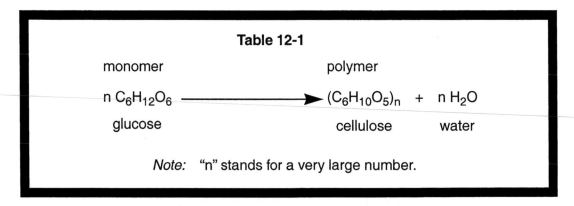

Table 12-1

monomer	polymer
n $C_6H_{12}O_6$ $\longrightarrow$	$(C_6H_{10}O_5)_n$ + n H_2O
glucose	cellulose water

Note: "n" stands for a very large number.

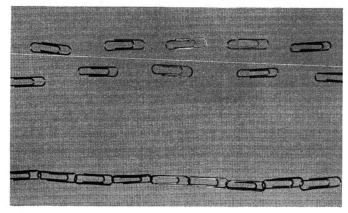

FIGURE 12-1
Each paper clip is a unit (mer). They can be linked together to make long chains of many units (polymers).

12-2a   What Kinds of Polymers Are There?

There are thermoplastic polymers, thermosetting polymers, and elastomers. Each type has different properties and uses.

12-2b   What Are Thermoplastic Polymers?

Thermoplastic polymers will melt and remelt every time they are heated. They can be reshaped many times. Polyvinylchloride (PVC), nylon, lucite, polystyrene, and saran are all examples of thermoplastic substances. When the plastic is cold, the long chain molecules are held in place, but when the plastic is heated, the long chains can slide over each other and change their shape.

12-2c   What Are Thermosetting Polymers?

Thermosetting plastics will not soften again once they have set. If they are subjected to intense heat, they will char. These plastics must be molded into their final

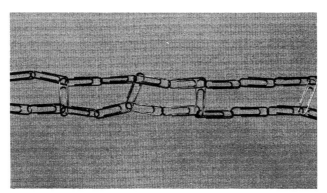

FIGURE 12-2
Thermosetting plastics will not soften again after they set. The giant molecular chains cross-link so that they cannot move across each other.

shape. Their shapes can be changed if they are filed or machined. **Bakelite** is an example of a thermosetting plastic. It is resistant to electricity and high temperatures. It is useful as electrical insulators and pot handles. Under proper conditions, some plastics can set at room temperature. These are called **cold-setting** plastics.

12-2d   What Are Elastomers?

Elastomers are polymers that have a high degree of elasticity. Rubber is a natural elastomer. This polymer was discovered by the Spanish explorers when they came to the New World. The Indians of South America coated their feet by dipping them into the latex (sap) of the rubber tree (Hevea brasiliensis). They wore this coating as shoes. This white sap has a mild odor which becomes stronger in a short time. Natural rubber becomes sticky in warm weather, and brittle at cold temperatures. It has limited usefulness. The famous English scientist, Michael Faraday, discovered that this polymer could be used to rub pencil marks from paper and therefore named it "rubber." In 1839, Charles Goodyear mixed sulfur with rubber and accidentally left it near his stove. When it cooled, the rubber hardened and become tough. He called the process of hardening rubber with sulfur **vulcanization**. The Firestone Tire Company is named after sulfur, which is also known as brimstone, or burning stone.

Elastomers have long, folded molecules that act as molecular springs. Energy is needed to stretch the molecules out. This energy can be felt as heat. Stretch a rubber band against your lips. You will feel the heat of friction as the molecular parts rub against each other. **Elasticity** is the ability of a substance to return to its original shape after being deformed.

12-2e   What Are Fibers?

We have natural and synthetic fibers. Natural fibers are derived from either animal or plant sources. Vegetable fibers are carbohydrate polymers $(C_6H_{10}O_5)_n$. Animal fibers are polymers of amino acids (proteins). Fibers made by the chemist resemble both groups of natural fibers. Under the microscope, natural fibers are not smooth. Synthetic fibers are smooth, and can be woven into strands for added strength. After stretching, synthetic strands actually increase in strength because the individual fibers fall more closely together. Pound for pound, nylon is stronger than steel cable! If nylon were not vulnerable to solvents, burning and softening when heated, it could replace steel in the construction of giant lightweight bridges. In some cases, automobile gears are made of nylon instead of steel because nylon has better wearing properties. "Composite" or plastic materials are being used more now in the manufacture of aircraft and automobiles because of their lower costs and higher performance.

Courtesy: Goodyear Tire and Rubber Co.

FIGURE 12-3a Charles Goodyear discovered how to vulcanize rubber in 1839.

FIGURE 12-3b

ALTERNATE ROUTE—A tapper at a rubber plantation on the Indonesian island of Sumatra uses an extension knife to draw latex from a rubber tree. The bark of a rubber tree is cut up part of the year and down the rest, to allow the tree to replenish itself. The rubber is sold to manufacturers for making such diverse products as surgical gloves, balloons, overshoes, and carpet backing.

FIGURE 12-3c

HARD DAY'S WORK—The rising sun lights the way as a tapper at the Dolok Merangir plantation in Sumatra deftly cuts a rubber tree for creamy latex to be used in radial tires, baby-bottle nipples, and other rubber products. The tapper will leave his mark on as many as 500 trees before he completes his duties in the early afternoon.

Courtesy: Goodyear Tire and Rubber Co.

FIGURE 12-3d

The manufacture of automobile tires.

FIGURE 12-3e

CLIMB TO THE TOP—A spiral drying tower carries crumbs of synthetic rubber to an inspector. After drying, the crumbs are compressed into bales and shipped around the world to manufacturers of everything from baby-bottle nipples and shoe soles to underwear elastic and auto tires.

FIGURE 12-3f

SMALL BUT MIGHTY— An air spring identical to 44 others used in a suspension system to simulate flight conditions for the space shuttle, *Challenger*.

Courtesy: Goodyear Tire and Rubber Co.

Natural and synthetic fibers magnified 300 times. Notice that the synthetic (plastic) fiber is more smooth than the natural fiber. Which fiber would hold dirt?

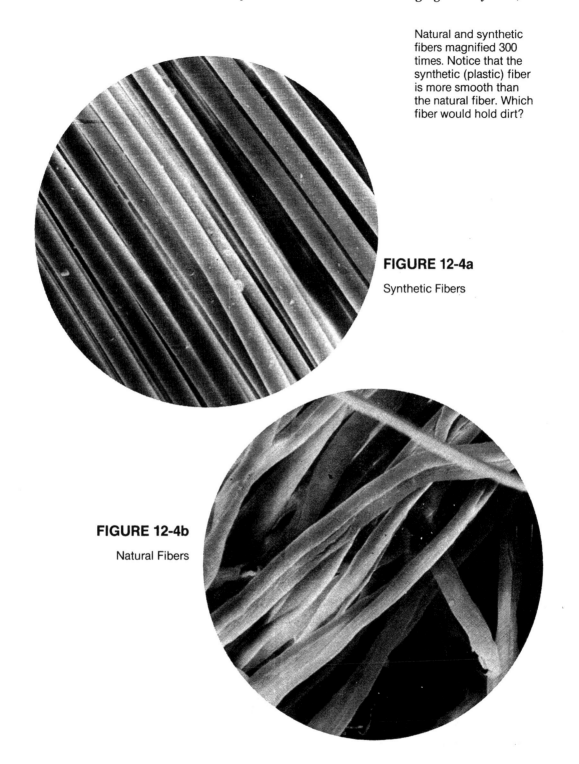

FIGURE 12-4a

Synthetic Fibers

FIGURE 12-4b

Natural Fibers

12-3a   How Are Polymers Formed?

Monomers are joined together through the process of polymerization. This process takes place by addition reactions or by condensation reactions.

12-3b   How Are Polymers Made in Addition Reactions?

Addition polymerization is the linking of small monomers. Double bonds between carbon atoms open up and join together in the addition process. They form almost endless chains. (See section 10-7 and Fig. 12-1.)

Table 12-2A   Ethylene Monomers

$$\cdots \ + \ \overset{\displaystyle H}{\underset{\displaystyle H}{C}} = \overset{\displaystyle H}{\underset{\displaystyle H}{C} } \ + \ \overset{\displaystyle H}{\underset{\displaystyle H}{C}} = \overset{\displaystyle H}{\underset{\displaystyle H}{C}} \ + \ \overset{\displaystyle H}{\underset{\displaystyle H}{C}} = \overset{\displaystyle H}{\underset{\displaystyle H}{C}} \ + \ \overset{\displaystyle H}{\underset{\displaystyle H}{C}} = \overset{\displaystyle H}{\underset{\displaystyle H}{C}} \ + \ \overset{\displaystyle H}{\underset{\displaystyle H}{C}} = \overset{\displaystyle H}{\underset{\displaystyle H}{C}} \ + \ \cdots$$

Table 12-2B

$$\cdots -\!\!\overset{H}{\underset{H}{C}}\!-\!\overset{H}{\underset{H}{C}}\!-\!\overset{H}{\underset{H}{C}}\!-\!\overset{H}{\underset{H}{C}}\!-\!\overset{H}{\underset{H}{C}}\!-\!\overset{H}{\underset{H}{C}}\!-\!\overset{H}{\underset{H}{C}}\!-\!\overset{H}{\underset{H}{C}}\!-\!\overset{H}{\underset{H}{C}}\!-\!\overset{H}{\underset{H}{C}}\!-\!\overset{H}{\underset{H}{C}}\!-\!\overset{H}{\underset{H}{C}}\!-\!\overset{H}{\underset{H}{C}}\!-\!\overset{H}{\underset{H}{C}}\!-\!\overset{H}{\underset{H}{C}}\!-\!\overset{H}{\underset{H}{C}}\!-\!\overset{H}{\underset{H}{C}}\!-\!\overset{H}{\underset{H}{C}}\!-\!\overset{H}{\underset{H}{C}}\!-\cdots$$

Polyethylene $(C_2H_4)_n$ is formed by the addition of ethylene monomers. The clear plastic bags used by vegetable markets and dry cleaners are made of polyethylene.

If the hydrogen is replaced with other elements, or groups of elements, new polymers will be created. Teflon is a polymer of tetrafluoroethylene (C_2F_4). Teflon is tough and has a greasy feel to its surface. Greaseless frying pans are coated with teflon. Foods do not stick to it when they are fried with little or no oil. Teflon will retain its properties over a wide temperature range ($-450°F$ to $+500°F$). It is used for insulation and tubing on jet aircraft, where extremes of temperature are encountered. Teflon is also used to line chemical tanks that hold acids, bases, and solvents. It is a space age polymer.

Table 12-3

$$\overset{F}{\underset{F}{C}} = \overset{F}{\underset{F}{C}} \ + \ \overset{F}{\underset{F}{C}} = \overset{F}{\underset{F}{C}} \ \longrightarrow \ -\!\overset{F}{\underset{F}{C}}\!-\!\overset{F}{\underset{F}{C}}\!-\!\overset{F}{\underset{F}{C}}\!-\!\overset{F}{\underset{F}{C}}\!-\!\overset{F}{\underset{F}{C}}\!-\!\overset{F}{\underset{F}{C}}\!-\!\overset{F}{\underset{F}{C}}\!-\!\overset{F}{\underset{F}{C}}\!-\!\overset{F}{\underset{F}{C}}\!-$$

tetrafluoroethylene   teflon

Table 12-4

Show how each of the monomers can add to form polymers.

Monomer	Polymer	Use
ethylene $\begin{array}{cc} H & H \\ \| & \| \\ C = C \\ \| & \| \\ H & H \end{array}$	polyethylene	toys plastic bags
chloroprene $\begin{array}{c} H \quad H \qquad\qquad H \\ \| \quad \| \qquad\qquad \\ C = C - C = C \\ \| \qquad \| \qquad H \\ H \qquad Cl \end{array}$	neoprene synthetic rubber	rubber products gasoline hoses
methyl methacrylate $CH_2 = C$ with CH_3 and $C-O-CH_3$ ($\underset{O}{\overset{\|\|}{\,}}$)	polymethylmethacrylate	substitute for glass, and use in aircraft
styrene $\begin{array}{cc} H & H \\ \| & \| \\ H = C \\ \| \\ H \end{array}$ ⬡	polystyrene	molds for the packaging of items hot coffee cups
propylene $\begin{array}{ccc} H & H & H \\ \| & \| & \| \\ C = C - C - H \\ \| & & \| \\ H & & H \end{array}$	polypropylene	fibers, and plastic hinges
dichloroethylene $\begin{array}{cc} H & Cl \\ \| & \| \\ C = C \\ \| & \| \\ H & Cl \end{array}$	saran	clear plastic food wrap
vinyl chloride $\begin{array}{cc} H & H \\ \| & \| \\ C = C \\ \| & \| \\ H & Cl \end{array}$	polyvinylchloride (PVC)	film, inflatable toys plastic pipe

12-3c   How Is Synthetic Rubber Made?

Synthetic rubber is the product of an addition reaction. It was first made by German chemists in World War I. This development was continued during World War II when the United States was cut off from its natural rubber sources in Southeast Asia by the Japanese. American chemists developed GR-S (government rubber-styrene), which is in use today. This rubber has better wearing qualities than natural rubber.

Table 12-5

butadiene + styrene ⟶ GR-S synthetic rubber

12-3d   How Are Polymers Made by Condensation?

When monomers combine by condensation, small molecules (H_2O and HCl) are formed and released. Nylon and Bakelite are formed in this way. Bakelite was invented by a brilliant Belgian chemist, Leo Bakelite, in 1909. He was attempting to find a substitute for shellac. Bakelite cannot rust, chip, or burn. It is used to make telephones, radio cabinets, ashtrays, combs, fountain pens, and electric insulators.

Table 12-6

formaldehyde + phenol ——CONDENSATION⟶ Bakelite + water

Nylon monomers are joined together through nitrogen atoms. They form an amide linkage (R—NH—R″). It is similar to that of proteins. Nylons and proteins are polyamides.

Table 12-7

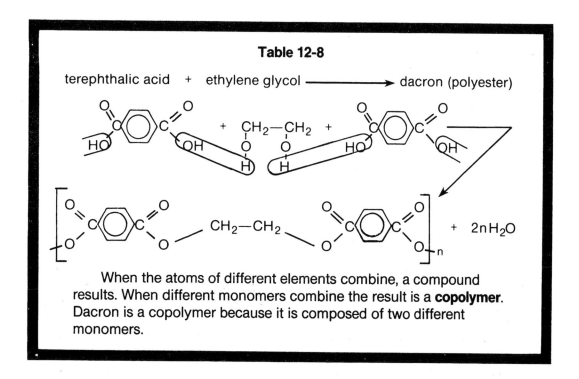

hexamethylenediamine + sebacylchloride ──→ 6-10 nylon + nHCl

Nylon is formed by condensation reactions. Cellulose is a condensation polymer of the monomer $C_6H_{12}O_6$ (see Table 12-1).

Table 12-8

terephthalic acid + ethylene glycol ──────→ dacron (polyester)

When the atoms of different elements combine, a compound results. When different monomers combine the result is a **copolymer**. Dacron is a copolymer because it is composed of two different monomers.

12-4   What Are Silicones?

Silicones are short chains of silicon and oxygen atoms with organic groups bonded to the silicon atoms.

Table 12-9

$$\ldots-O-\underset{\underset{R}{|}}{\overset{\overset{R}{|}}{Si}}-O-\underset{\underset{R}{|}}{\overset{\overset{R}{|}}{Si}}-O-\underset{\underset{R}{|}}{\overset{\overset{R}{|}}{Si}}-O-\underset{\underset{R}{|}}{\overset{\overset{R}{|}}{Si}}-O-\ldots$$

R represents an organic radical.

"Silly Putty," and "Super Ball" are made of silicones. Short-chain silicones are useful lubricants. They are oil substitutes that do not change their viscosity (thickness) over a wide temperature range (–100°F to +400°F). Silicone rubbers maintain their elasticity over a greater temperature range than any natural rubber (–130°F to +600°F). Silicones have wide applications, from space vehicles to coatings for automobiles.

12-5　How Can Synthetic Polymers Be Disposed Of?

Each year more than one million tons of plastics are buried in landfills or dumped into the oceans. The plastics float in the water and are navigational hazards for ships. Very often, floating plastics may look like food to marine creatures. These plastics cannot be digested and may cause the death of the fish or animal. At times, fish and marine animals become entangled in discarded nylon fishing lines. Discarded plastics may take as long as fifty years to disintegrate. In the meantime, we continue to discard plastics, and the debris continues to grow.

There are problems with the incineration of polymers. When chlorinated polymers are incinerated, hydrogen chloride is produced. This poisonous gas will kill people and corrode buildings and metallic objects. Vinyl cyanides, like orlon, burn to produce lethal (deadly) hydrogen cyanide gas. Perhaps incineration at sea (see section 13-9d) can resolve this problem.

Chemists are developing photodegradable polymers (polymers that are broken down by sunlight). This can be done by inserting light-sensitive chemical groups at regular intervals into the polymer chain. The light (energy) is absorbed by the chemical groups and breaks the large molecules apart. The small parts that are left are made biodegradable (able to be decomposed by microorganisms). These polymers will not disintegrate in artificial light, like fluorescent lighting. They are sensitive only to ultraviolet light, found in sunlight. Plastics discarded in the Arctic and Antarctic regions of the earth would continue to exist for long periods of time. Biodisintegration is very slow in cold climates. It is for this reason that prehistoric mammoths that died thousands of years ago are found preserved in the frozen ground of Siberia.

Now You Know

1.   The scientist copies from nature in developing chemical creations.

2.   The first synthetic plastic was celluloid. It made the movie industry possible.

3.   A polymer consists of enormous molecules made of many smaller molecular units called monomers. They are linked together in almost endless chains.

4.   There are natural polymers and those made by chemists.

5.   Natural polymers are derived from plants and animals.

6.   Polymers can be classified as thermoplastic, thermosetting, and elastomer.

7.   Thermoplastic polymers will soften repeatedly, for as many times as they are heated. Some examples of thermoplastic polymers are nylon and polyethylene.

8.   Thermosetting polymers will not soften again once they are molded. Bakelite is an example of a thermosetting polymer.

9.   Elastomers are polymers that regain their shape after being distorted (stretched). Rubber is a natural elastomer.

10. Synthetic fibers are stronger than natural fibers, hold less dirt, and are easier to clean. The microscope reveals that synthetic fibers are smoother than natural fibers.

11. Polymers are formed by addition reactions or by condensation reactions.

12. Polymers formed by addition reactions require monomers that are double bonded between carbon atoms.

13. Polymers formed by condensation reactions result in the splitting out of a small molecule, such as H_2O or HCl.

14. Copolymerization is polymerization in which different monomers unite.

15. The monomers of nylon and proteins are combined by amide linkages.

16. Silicones consist of chains of silicon and oxygen with organic radicals bonded to the silicon.

17. The disposal of nonbiodegradable polymers presents problems. They cannot be buried in sanitary landfills. They cannot be incinerated in populated areas because poisonous gases are produced when they burn.

New Words

addition polymerization	The combining of monomers into polymers.
amino acids	The molecular monomers that combine to form natural protein.
cold-setting plastic	A thermosetting plastic that will set at room temperature.
condensation polymerization	The combination of monomers to form polymers in which a small molecule, such as H_2O or HCl, is formed and released.
copolymer	A polymer that is made of different monomers.
elastomer	A polymer that can stretch or be distorted under the stress of a force. It will return to its original shape.
macromolecule	A giant molecule.
monomer	The small molecular units that join together to make a polymer.
plastic	A polymer.
polymer	"Many units" or many monomers that are joined together to make a giant molecule.
silicones	A chain of silicon and oxygen atoms. Carbon radicals are bonded to the silicon atoms.
thermoplastic polymer	A polymer that will soften each time that it is heated.
thermosetting polymer	A polymer that will not soften again after it sets.
vulcanization	The addition of sulfur to rubber to harden and toughen it.

Reading Power

For each of the following questions, select one answer that seems most correct.

1. The main idea of the introduction is:
   a. Celluloid was the first plastic ever made.
   b. The creative chemist copies from nature.
   c. Celluloid is very flammable.
   d. Celluloid made the movie industry possible.

2. The main idea of section 12-1 is:

   a. Cellulose was the first polymer made.
   b. There are natural and synthetic polymers.
   c. Plastics are giant macromolecules.
   d. Nylon is a polymer that is stronger than steel.

3. The main idea of section 12-2 is:

   a. Thermoplastic polymers will soften each time that they are heated.
   b. Thermosetting plastics will not soften again after they are set.
   c. Elastomers will return to their original shape after they are distorted.
   d. Plastics may be thermoplastic, thermosetting, or elastomers.

4. The main idea of section 12-3a is:

   a. Monomers form polymers by addition and condensation reactions.
   b. Monomers form polymers by addition reactions.
   c. Monomers form polymers by condensation reactions.
   d. Monomers form polymers.

5. The main idea of section 12-3b is:

   a. A copolymer is made by the addition of different monomers.
   b. Monomers that are useful in the addition reaction are all single bonded.
   c. Addition polymerization takes place at the double bond between the carbon atoms.
   d. Addition polymerization takes place when water molecules are formed.

6. The main idea of section 12-3c is:

   a. Synthetic rubber was first made by German chemists.
   b. Synthetic rubber is formed by an addition reaction.
   c. The basic materials for synthetic rubber are butadiene and styrene.
   d. All of the above.

7. The main idea of section 12-3d is:

   a. Condensation polymerization is the combining of monomers in which small molecules are formed and released.
   b. Bakelite is a thermosetting plastic.
   c. Nylon is a polyamide.
   d. Cellulose is a condensation polymer.

8. The main idea of section 12-4 is:

   a. Silicones have properties that show little change over very wide temperature ranges.
   b. Silicones are chains of silicon and oxygen atoms.
   c. Short silicone chains are liquids that are used as lubricants.
   d. All of the above.

9. The main idea of section 12-5 is:
   a. Synthetic polymers are difficult to dispose of.
   b. Synthetic polymers are generally not biodegradable.
   c. Synthetic polymers must be buried in the ground.
   d. Natural polymers are biodegradable.

Mind Expanders

1. Why are synthetic fibers easier to clean than natural fibers?

2. How are cotton and wool different? How are they the same?

3. How are nylon and protein similar?

4. How are thermoplastic polymers different from thermosetting polymers and elastomers?

5. Since nylon is stronger than steel, why isn't nylon used to construct bridges and airplanes?

6. How is addition polymerization different from condensation polymerization?

7. What are the advantages of silicones over carbon polymers?

8. What are the problems involved with the disposal of synthetic polymers?

Completion Questions

Write the word(s), from the list below, that will correctly complete the following statements. You may use the same word(s) more than once.

rubber	condensation	addition
copolymer		the silicones

1. A polymer made from two different monomers is called a _____ .

2. An example of a natural elastomer is _____ .

3. A polymer such as _____ has properties that remain constant over a wide temperature range.

4. The two main chemical reactions by which polymers are formed are

   _____ and _____ .

5. The _____ reaction results in the splitting out of a water molecule from the uniting monomer units.

True or False

If the statement is true, mark it *T*. If the statement is false, change the indicated word(s), using the list below, to make the sentence true.

Bakelite	silicone	thermoplastic
natural fibers	incinerated	thermosetting
monomer	elastomer	copolymer

1. The first plastic, made by John W. Hyatt, was *celluloid.*

2. A *polymer* consists of giant molecules formed by the joining together of smaller units.

3. Plastics made of different monomers are called *copolymers.*

4. The process of adding sulfur to rubber to toughen it is called *vulcanization.*

5. A chain of silicon and oxygen, on which organic radicals are attached to the silicon atoms is called a *thermosetting polymer.*

6. *Elastomers* tend to be water-repellent and resistant to electricity.

7. Synthetic fibers may release poisonous gases when they are *buried.*

8. When a distorting force is applied to a *thermosetting plastic*, the molecules restore their original shape.

9. Natural fibers can be safely buried because they are *biodegradable.*

10. *Synthetic fibers* grow stronger with use.

Find the Facts

Locate the section in Chapter 12 that supports each answer to the true/false questions. Copy the sentence into your notebook near your answer.

Multiple-Choice

1. Addition polymerization can take place with each of the following monomers except:

   a. $CH_2\!=\!CH_2$ b. $CH_3\!-\!CH\!=\!CH_2$

   c. $CH_3\!-\!CH_3$ d. $CH(CN)\!=\!CH_2$

2. Teflon is a tough polymer formed by the addition of:

   a. $CH_2\!\!=\!\!CH_2$ b. $CF_2\!\!=\!\!CF_2$
   c. $CH(CN)\!\!=\!\!CH_2$ d. $CH_3\!\!-\!\!CH\!\!=\!\!CH_2$

3. The following is an example of a copolymer:

   a. dacron   b. polyethylene   c. teflon   d. polystyrene

4. A polymer that is similar to a protein is:

   a. nylon   b. polyethylene   c. teflon   d. saran

5. The following polymers are made in a condensation reaction except:

   a. Bakelite   b. nylon   c. dacron   d. synthetic rubber

Matching

Write the number in column B that relates to the word(s) in column A in the space indicated.

COLUMN A	COLUMN B
_____ thermoplastic	1. A polyamide stronger than steel.
_____ GR-S	2. Polyvinylchloride.
_____ nylon	3. Polyethylene.
_____ PVC	4. Synthetic rubber.

Polymer Creation

1. Using the table of monomers, draw the molecular structure of the polymer formed by the addition of each monomer with itself.

2. Draw the polymer formed by the addition of two different monomers (copolymers). How many different copolymers can you create?

Lab Exercise

Make ball and stick models of two monomers, then show how they polymerize. Your teacher may choose to put the best polymer models on display.

Chapter 13

What Is the Effect of Chemistry on Our Ecology?

Instructional Objectives

After completing this chapter, you will be able to:

1. Define aerosols, anaerobic bacteria, biodegradable, BOD, ecology, eutrophication, food chain, greenhouse effect, pheremones, pollution, and saltation.

2. Recognize the dynamic exchange that takes place between living things and our mineral environment.

3. Explain the role of each component of air.

4. Explain how various types of combustion occur and how fires can be extinguished.

5. Explain how natural and polluting acid rain forms.

6. Compare natural acid rain with acid rain pollution.

7. Define four air pollution problems and their possible remedies (e.g., acids, aerosols, automobile, and industrial pollutants).

8. Discuss four ways our water supply can be polluted and suggest possible remedies for each problem (detergents, metals, sewage, and pesticides).

9. Discuss three ways our land can be polluted and suggest possible remedies for each problem (highway construction, mining, chemical dumping).

Chapter 13 Contents

	Introduction	311
13-1	What Is "Normal" Air?	313
13-2	How Does the Air Support Combustion?	313
13-3	What Are Some Types of Combustion?	313
13-4	How Can Fires Be Extinguished?	315
13-5a	How Are Rains Formed?	315
13-5b	How Is Carbonic Acid Formed in Rain?	315
13-5c	How Is Nitric Acid Formed in the Air?	316
13-5d	How Are Sulfuric Acid Rains Formed?	316
13-6a	How Does Nature Change Our Air?	317
13-6b	How Does Dust Get into Our Air?	317
13-6c	How Are Other Aerosols Suspended in Our Air?	317
13-7a	How Can Carbon Dioxide Pollute Our Air?	317
13-7b	How Does Carbon Monoxide Pollute Our Air?	318
13-7c	How Does Sulfur Dioxide Pollute Our Air?	321
13-7d	How Does the Automobile Contribute to Air Pollution?	321
13-7e	How Can Alternate Fuels Power Our Automobiles?	322
13-7f	How Does Industry Pollute the Air?	323
13-8a	How Safe Is Our Water Supply?	325
13-8b	How Do Detergents Contribute to Water Pollution?	325
13-8c	How Do Metals Contribute to Water Pollution?	329
13-8d	How Is Sewage Treated?	329
13-8e	How Do Pesticides Contribute to Water Pollution?	330
13-9a	How Do Humans Pollute the Land?	330
13-9b	How Do Humans Exploit and Destroy the Topsoil?	330
13-9c	How Do Solid Wastes Pollute Our Land?	330
13-9d	How Are Solid Wastes Incinerated?	331
13-9e	How Do Sanitary Landfills Solve the Problems of Solid Waste Disposal?	331
13-9f	How Else Have Humans Polluted the Land?	331
13-10	How Can Poisonous Chemicals Be Eliminated?	331
	Written Exercises	334
	Career Information	339

Chapter 13

What Is the Effect of Chemistry on Our Ecology?

Introduction

There is a spaceship orbiting the sun at a speed of 18.5 miles each second. This space vehicle is carrying more than 4.5 billion humans, in addition to billions of other life forms, in a balanced chemical environment. The living and nonliving contents of this spaceship are all part of a vast chemical system. The life forms draw needed chemicals from the nonliving parts of the orbiting spacecraft, and repay the chemicals as fast as they are drawn. The system is balanced by the equal exchange of chemicals between the living and nonliving components. The entire system is powered by energy from the sun. Recently problems began to emerge on this spaceship. One life form developed a high level of technology in its quest to improve its life-style. It began to draw more chemicals than is natural and to create new chemicals that were strange to the environment. It was found that when one part of the environment was disturbed, it affected all other components, both living and nonliving. The spaceship described is our planet, the earth. In this chapter, we will find out how people are changing their chemical environment. Ecology is the study of the relationship between living things and the environment.

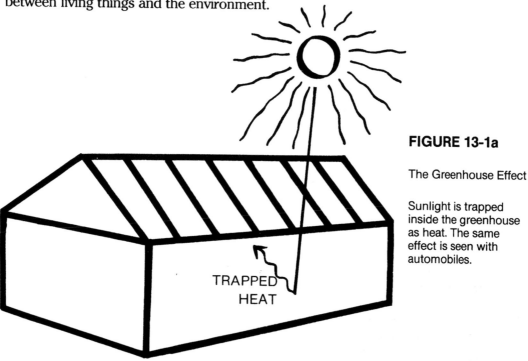

FIGURE 13-1a

The Greenhouse Effect

Sunlight is trapped inside the greenhouse as heat. The same effect is seen with automobiles.

TRAPPED HEAT

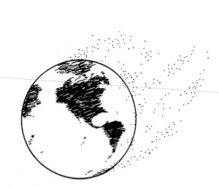

FIGURE 13-1c

The earth in orbit, attracting cosmic dust as it sweeps a path through space.

FIGURE 13-1d

Industry and some power-generating plants spew pollutants into our atmosphere.

FIGURE 13-1b

The Mt. St. Helens volcano erupting. Volcanoes are major contributors to atmospheric dust.

Courtesy: Goodyear Tire and Rubber Co.

13-1   What Is "Normal" Air?

Air is a mixture of solids, liquids, and gases. The air is about 79% nitrogen and 20% oxygen. The remaining 1% comprises dust (solid), water (vapor and droplets), carbon dioxide, argon, neon, ozone, and other trace gases. Some solids (and liquids) exist as microscopic particles or **aerosols**. They are formed naturally from volcanoes, forest fires, and from the sweep of the earth through space. Space is relatively "dirty," and the earth, as it travels around the sun, attracts much dust with its gravity. Dust is an important component of our air. Without dust, there could be no clouds, rain, or snow. Agriculture, even civilization as we know it, would probably not exist. The land portion of the earth would most likely become a giant desert. Atmospheric dust accounts for beautiful sunsets and the haze in the air. Carbon dioxide is generated from the burning of all carbon fuels and the exhaling of all animals.

13-2   How Does the Air Support Combustion?

The ancient Greeks believed that burning substances simply lost the element "fire" (see section 3-1). In 1687, German chemist Georg Ernst Stahl (1660–1734) said that when things burn, they lose "phlogiston" (*phlogiston* is the Greek word that means "to set on fire"). According to this theory, burning objects should lose weight, since they lose their phlogiston. In 1780, Antoine Lavoisier (1743–1794), as a result of careful measurements, overturned the phlogiston theory of burning. He found that when metals burn, a *gain* in weight results. When wood burns, the weight of its products (ashes, carbon dioxide, and water) is greater than that of the unburned wood. By careful experimentation, he discovered that all burning needs oxygen from the air. If you place a jar over a burning candle, the flame will go out once the oxygen in the air is consumed. All living things need a constant supply of oxygen to burn food in their cells. Lavoisier showed that *all burning requires oxygen, a fuel, and an ignition (or kindling) temperature.* The ignition temperature is the lowest temperature at which burning will begin. Lavoisier is remembered as the "father of modern chemistry." Unfortunately, he was guillotined during the French Revolution because he was a member of the aristocracy.

13-3   What Are Some Types of Combustion?

Combustion may be rapid or slow, complete or incomplete, spontaneous, or explosive.

Rapid oxidation is characterized by the release of heat or both heat and light. Slow oxidation shows very little heat, and no light. Paper and gasoline burn rapidly, but iron rusts slowly.

If a hydrocarbon fuel (see section 10-3) is burned in an abundance of oxygen, carbon dioxide and water will result. Maximum heat is generated, and there will not be any ashes. Methane (CH_4) is the common gas used on home stoves. It burns as follows:

$$CH_4 + 2\,O_2 \xrightarrow[\text{combustion}]{\text{complete}} CO_2 \uparrow + 2\,H_2O + \text{heat}$$

methane   oxygen   carbon   water
                                      dioxide

Care should be taken when burning natural gas at home to see that all air holes are clear. If the air is blocked, the methane may not burn completely, and deadly carbon monoxide may be produced.

$$2\,CH_4 + 3\,O_2 \xrightarrow[\text{combustion}]{\text{incomplete}} 2\,CO \uparrow + 4\,H_2O + \text{heat}$$

methane   oxygen   carbon   water
           (limited)   monoxide

A yellow sooty flame on your stove indicates that a limited amount of air is mixing with the stove gas. Your fuel bills will be higher, your pots and pans may show a black carbon deposit, and you will cook with a cool flame. A yellow flame indicates unburned carbon. The problem can be solved by cleaning the air inlets.

$$CH_4 + O_2 \xrightarrow[\text{combustion}]{\text{incomplete}} C \text{ (soot)} + 2\,H_2O$$

methane   oxygen   carbon   water
           (limited)

Under certain conditions, fires can break out "by themselves." This **spontaneous combustion** can take place in your home in a pile of oily rags left in a closed space such as a closet. In factories and mines, fires have broken out in large piles of coal. On farms, fire can break out spontaneously in bales of hay. Decay organisms generate heat, which builds up and is stored, until the ignition temperature is reached. This slow buildup may take weeks or months. At that point, rapid combustion will take place. Spontaneous combustion is rapid burning due to the accumulation of heat from slow oxidation.

An explosion is a very rapid and uncontrolled reaction. Explosive mixtures of gasoline and air are used in automobile engines to generate motion. Dust explosions take place if a spark is introduced into small, fine particles suspended in the air. Dust explosions can occur in flour mills. Small particles present large reacting surfaces. When contact between chemicals increases, reactions take place more rapidly (see section 5-12).

13-4   How Can Fires Be Extinguished?

Fires can be prevented if they are deprived of fuel, oxygen, or the necessary ignition temperature. Care and good sense must be used to avoid destructive fires. When electrical appliances are in use, they should not be left unattended. Cooking in the kitchen demands constant attention.

Fires can be extinguished by several methods. Smothering the fire with a wet blanket or sand will deprive the fire of needed oxygen. Carbon dioxide fire extinguishers also serve to put out fires. "Firefoam" is used on burning liquids such as oil. *Never* throw water on an oil fire. Water is most commonly used to quench ordinary fires. It serves to bring the temperature of the burning fuel below the ignition point. Very often, the best way to put out a fierce fire is to remove its fuel. Many forest fires are stopped by digging out the vegetation in front of the fire.

Fires pollute our air by pouring poisonous gases and excessive dust into the atmosphere.

13-5a   How Are Rains Formed?

Rainfall takes place when water droplets, suspended in the air as clouds, grow heavy enough to fall to the ground. A cloud is a suspension of liquid water droplets (or ice crystals) in the air. A fog is a cloud at ground level. When water evaporates from our rivers, lakes, and oceans, the vapor rises on the warm air currents. At higher altitudes, the vapor condenses or sublimates on cool specks of dust. There is a speck of dust in each raindrop and snowflake. When it rains, the air is cleansed of its suspended dust. Green plants need rain. With no dust in the air, there could be no rain (or snow). The earth would become a great desert with no rivers or lakes.

13-5b   How Is Carbonic Acid Formed in Rain?

Carbon dioxide in the air dissolves in the rain, forming a dilute carbonic acid solution.

Table 13-1

$$CO_2 + H_2O \longrightarrow H_2CO_3$$
carbonic acid

$$H_2CO_3 + H_2O \rightleftharpoons HCO_3^- + H_3O^+$$
carbonic acid

The carbonic acid in our rain forms from carbon dioxide that dissolves in the rainwater.

13-5c   How Is Nitric Acid Formed in the Air?

During lightning storms, the nitrogen in the air combines with atmospheric oxygen to form oxides of nitrogen. These oxides continue their oxidation to nitric acid. The acid dissolves in the rainwater. Nitric acid is a natural component in rainwater during lightning storms. This is important in providing plants with nitrates, a natural fertilizer. Nitrates are needed by green plants to make proteins.

Table 13-2

$$N_2 + O_2 \text{ —— LIGHTNING ——} 2\,NO \text{ (nitric oxide)}$$

$$2\,NO + O_2 \longrightarrow 2\,NO_2 \text{ (nitrogen dioxide)}$$

$$3\,NO_2 + H_2O \longrightarrow NO + 2\,HNO_3 \text{ (nitric acid)}$$

13-5d   How Are Sulfuric Acid Rains Formed?

In the Bible, there are many references to brimstone, or burning stone. Brimstone is the element sulfur. This element will burn to form the colorless and choking gas, sulfur dioxide. Continued oxidation of this gas results in sulfuric acid. Sulfur dioxide is produced naturally by volcanoes.

Table 13-3

$$S + O_2 \longrightarrow SO_2 \text{ (sulfur dioxide)}$$

$$2\,SO_2 + O_2 \longrightarrow 2\,SO_3 \text{ (sulfur trioxide)}$$

$$SO_3 + H_2O \longrightarrow H_2SO_4 \text{ (sulfuric acid)}$$

The concentration of sulfuric acid at pollution levels is more than ten times greater than natural rain (the pH of natural rainfall is 5.5, while the pH of polluted acid rain is about 4.2). It is caused by the outpouring of oxides of sulfur into the air, from the burning of sulfur-containing fuels, and the exposure of sulfur to the air, when coal is mined. Acid rain pollution is destructive to our forests and lakes.

13-6a   How Does Nature Change Our Air?

Acid rains that occur naturally benefit green plants by dissolving ground minerals. Normal air also contains solids in the form of dust, soil, and salt. It contains liquids in the form of water droplets.

13-6b   How Does Dust Get into Our Air?

Space is a major contributor of dust to our atmosphere. Scientists estimate that 31,700 tons of dust are swept into our atmosphere each second! Most of this is **cosmic dust**. Humans contribute about 15.85 tons of dust each second. About five out of every 10,000 parts of the earth's atmospheric dust comes from humans. Volcanoes are another major contributor of atmospheric dust. One eruption can release more than 100 billion cubic yards of dust into the air. These particles may rise fourteen miles and take many years to settle out. This blanket of solid material eventually surrounds the earth, and may cause a cooling of the climate. A new ice age could result if many severe volcanic eruptions were to occur. Suspended volcanic dust in the air produces beautiful deep blue and lavender sunrises and sunsets.

13-6c   How Are Other Aerosols Suspended in Our Air?

Other natural ways solids may be introduced into the air are by forest fires, dust storms, and sand storms. Soil becomes windblown when there is a lack of rain and plants. The process by which soil becomes windblown is called **saltation**. At the seashore, another pollutant, in the form of salt, creates a corrosion problem for automobile owners. The waves, breaking on the shore, throw seawater into the air. The wind may lift these salty droplets and carry the salt particles inland for short distances. Salt particles are not able to penetrate deeply into the land.

13-7a   How Can Carbon Dioxide Pollute Our Air?

Carbon dioxide is poured into the air from burning coal, oil, wood, and all other carbon fuels. The amount of carbon dioxide in the air has been increasing steadily since the industrial revolution began in the early 1800s. In 1890, the CO_2 level in the air was about 290 parts per million. In 1960, the concentration of CO_2 in the air increased to about 315 parts per million. If the concentration of this gas were to double, the temperature of the earth would increase about six Fahrenheit degrees (6 F°). In addition, tiny amounts of new gases are entering our atmosphere (methane, ozone, and chlorofluorocarbons). They may also raise the temperature of the air by the **greenhouse effect**, or the trapping of heat by gases in the air. Today 15% of the solar energy reaching the earth is absorbed by our atmosphere (see Fig. 7-4). This energy is changed to heat. Carbon dioxide absorbs heat efficiently and will not allow the heat to escape back into space. A similar effect is found when sunlight shines into an automobile with closed windows. The sunlight is converted into heat and becomes trapped inside the car because it cannot escape through the closed windows. If the greenhouse effect should become a reality, the ice caps at the north and south poles would melt. The coastal cities would become flooded. The shorelines would be moved back to higher ground. In New York City, Manhattan, would be transformed into the "Venice of North America." There would be less land on the earth. The chances of this happening are being studied by scientists at this time. It is possible that the oceans may absorb the heat and protect us from drastic climatic changes. The ice at the poles may be able to absorb some heat without melting, if the ice remains below its freezing point. Just how serious carbon dioxide pollution will become is unknown.

FIGURE 13-2a

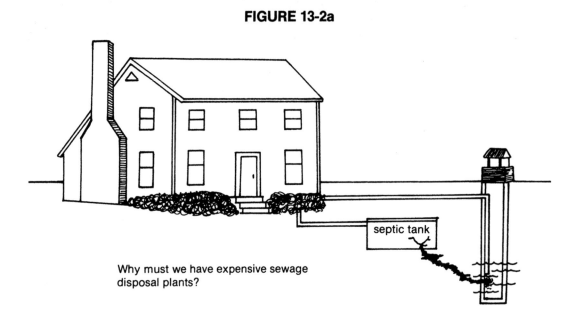

septic tank

Why must we have expensive sewage disposal plants?

13-7b   How Does Carbon Monoxide Pollute Our Air?

Carbon monoxide (CO) is a dangerous pollutant that has killed many people. This gas is colorless (invisible) and odorless. It is formed when a carbon-type fuel is incompletely burned. Incomplete burning takes place when there is a limited amount of air available. Automobiles are responsible for most carbon monoxide pollution. It is estimated that each automobile produces an average of five pounds of this gas each day.

Table 13-4

$$2\,C \;+\; O_2 \xrightarrow{\text{LIMITED AIR}} 2\,CO$$

FIGURE 13-2b

How can we dispose of
chemical wastes?

FIGURE 13-2c

Carbon monoxide is also a product of cigarette smoking. When carbon monoxide is inhaled into your lungs, it is absorbed into your blood. It combines with a protein (hemoglobin) in your red blood cells. As a result, the red cells can no longer carry oxygen to your tissues, including your heart. Since there is less usable blood for your body, there is an increased strain on the heart. Low doses of carbon monoxide may result in higher blood pressure, headache, and upset stomach. Death can result by suffocation. The victim appears to have very red lips and rich color in the face, because the blood takes on a bright red hue. The amount of carbon monoxide released by automobiles is declining because of the use of catalytic converters. They are required on all cars sold in the United States.

13-7c How Does Sulfur Dioxide (SO$_2$) Pollute Our Air?

Sulfur dioxide is an industrial waste that is produced wherever sulfur is burned. It is generated when coal is burned. High-sulfur oil also contributes a substantial share of sulfur dioxide to the air. The smelting of sulfide ores produces substantial amounts of this gas (see Chapter 9). It is a colorless gas that has a choking odor. It will irritate your eyes and lungs. Sulfur dioxide is changed into sulfuric acid in the air (section 13-5d). This acid is responsible for the death of fish in lakes hundreds of miles from the place where the sulfur is burned. Effective removal of sulfur dioxide from the smoke before it pours into the air is difficult. Since the cleanup of coal stack emissions, larger pieces of fly ash (unburned pieces of ash that are easily blown by the wind) have been eliminated. As a result, sulfur dioxide pollution has increased. This is because large pieces of ash are alkaline. They neutralize the acidic sulfur compounds before they can escape into the atmosphere.

13-7d How Does the Automobile Contribute to Air Pollution?

Americans love their automobiles. The carmakers produce autos that give their owners more miles to the gallon of gasoline. All cars are equipped with **catalytic converters**. These converters cut down on poisonous emissions from cars, like carbon monoxide, oxides of nitrogen, and hydrocarbons. In spite of all these improvements, automobiles remain the greatest single source of pollution in U.S. society, and they will be an even greater problem in the future.

American cars emit about as much carbon dioxide as is emitted in the industrial country of Japan. Cars are about 14% efficient. That means that only 14 gallons out of every 100 gallons of gasoline move the car forward. The remaining 86 gallons pollute our air, are lost as heat, and contribute to the greenhouse effect. More people are driving more cars more miles each year. In 1970, there were 2.5 Americans for each car in the United States. In 1990, there were 1.7 Americans per car. In 1970, Americans drove enough miles to make two million round-trips to the moon. In 1990, the mileage driven would come to three million round-trips. By the year 2000, the miles driven may increase to 4.2 million round-trips. As a result of the increasing mileage driven, the increased stop-and-go traffic, poisonous emissions are increasing.

Even with the big price increases due to the 1990 Persian Gulf conflict, the cost of gasoline is nearly the same as it was in 1970, after prices are corrected for inflation. More people live outside the large cities and must travel longer distances to work. People also enjoy driving. Cars are made to be very comfortable so that drivers can endure traffic jams more easily.

Oil imported to our refineries often spills in our waterways and land. The damage is costly to clean up and destructive to birds and fish. Automobiles require large imports of oil from foreign countries and the outflow of monies. In 1990, we paid the costs of 40 million fill-ups each day to other countries.

More automobiles are equipped with air conditioners. These devices encourage people to do more driving in greater comfort. Air conditioners lower gasoline mileage and are major contributors to the greenhouse effect. Air conditioners often leak their chlorofluorocarbon coolants into the air.

Air pollution from the automobile extends beyond carbon monoxide. High-compression engines produce oxides of nitrogen. Nitric oxide can form nitric acid (HNO_3). (See Table 13-2.)

Lead is added to gasoline to improve burning in the automobile engine. A bromine compound must also be added to gasoline to eliminate the possibility of a lead buildup in the engine. The lead is finally expelled from the tailpipe of the car as lead (IV) bromide ($PbBr_4$). This compound is a toxic gas, which can cause lead poisoning if inhaled. Leaded gasoline is slowly being phased out to eliminate this lethal pollutant. Lead-free gasoline is being used in its place. Unfortunately, lead-free gasoline may present problems of its own. Compounds are added to it to improve its burnability in the auto engine. These compounds are **carcinogenic** (cancer-producing) chemicals. It is possible that the danger from lead is being replaced by a new danger, in the form of the products produced by unleaded gas.

13-7e   How Can Alternate Fuels Power Our Automobiles?

It is estimated that the world will "run out of gas" by the year 2030. By that time automobiles will be propelled by fuels other than gasoline. This may sound like welcome news, since gasoline causes smoke that contains unburned hydrocarbons, carbon monoxide, and oxides of nitrogen. Since 1968, automobiles have been equipped with catalytic converters so that the unburned hydrocarbons (fuel) are completely converted into carbon dioxide and harmless water vapor. The oxides of nitrogen are also trapped by these converters. These devices are about 97% efficient. Today, scientists and engineers are examining alternative fuels that may reduce our dependence on oil and produce less pollution.

Methanol: Engineers have experimented with the use of wood alcohol (methanol) in place of gasoline. Methanol is easily made from natural gas or coal. It is poisonous. It can enter your body by absorption through the skin, or when you breathe its vapor. When it burns, it produces about 10% less carbon dioxide than gasoline. A major problem with methanol is that it produces about half of the energy that an equal amount of gasoline produces. That means cars would need about twice as much methanol as gasoline to travel the same distance. In addition, methanol may generate formaldehyde (embalming fluid) and emit it from the automobile exhaust system. Methanol causes extensive corrosion and engine damage. Engineers find a mixture of 85% methanol and 15% gasoline to be most efficient.

Ethanol: If methanol is poisonous, then why not use ethanol? This is the alcohol that is found in wine, beer, and other alcoholic drinks. Vodka is simply ethanol (grain alcohol) and water. Ethanol has been blended with gasoline and used for many years to power automboiles in Brazil and in parts of the United States. A mixture of 10%

alcohol and 90% gasoline is called "gasohol." The biggest problem with ethanol is its high cost. Grain alcohol is currently obtained from fermentation (see section 10-12). It costs about three times as much as methanol.

Liquid Hydrogen: Hydrogen is the least-polluting fuel. If the hydrogen tank should leak, this light element would move rapidly up through the air and into space. The gravity of the earth is not great enough to hold hydrogen in our atmosphere. When it burns, harmless water vapor is released.

Liquid hydrogen must be maintained at very low temperatures ($-253°C$). The refrigeration unit would require nearly as much energy as the car would need. If this fuel were used as a compressed gas, it would need more space for storage.

Electricity: The electric automobile is the cleanest-running vehicle. These cars use electricity from batteries to run lightweight, powerful electric motors. There are now two major problems with this technology. The batteries we have today need so much space and weight that the car can only travel about 100 miles on each charge. The second problem is that the used batteries need a long time to recharge. It would be difficult to replace the heavy batteries with fresh ones at a "refueling" station.

While electrical vehicles do not pollute, electrical power stations do add harmful gases to our air. The pollution from these cars appears at the smokestacks of electrical power stations.

13-7f How Does Industry Pollute the Air?

In December 1984, more than 2,000 people were killed in Bhopal, India, because of the release of poisonous methyl isocyanate into the air. This chemical is used in the production of insecticides. In April 1985, a congressional study showed that the chemical industry routinely releases tens of thousands of tons of carcinogens and pollutants into the air every year. These chemicals include carbon tetrachloride (CCl_4), chloroform ($CHCl_3$), benzine (C_6H_6), and chlorine (Cl_2). This study showed that West Virginia, Texas, and Louisiana, where there are many oil refineries and petrochemical plants, suffered from the worst pollution levels. These states also have high cancer rates.

Table 13-5

Properties of Some Important Gases

Gas	Physical Properties	Chemical Properties
carbon dioxide	colorless 1½ times heavier than air odorless tasteless slightly soluble in water	weak acid poisonous
carbon monoxide	colorless weighs the same as air odorless tasteless slightly soluble in water	poisonous burns reducing agent
hydrogen	colorless (invisible) lightest element in the universe odorless tasteless almost insoluble in water	burns does not support combustion
oxygen	colorless (invisible) slightly heavier than air odorless tasteless slightly soluble in water	supports combustion does not burn
sulfur dioxide	colorless more than twice as heavy as air choking gas sour taste moderately soluble in water	poisonous weak acid

13-8a   How Safe Is Our Water Supply?

Water pollution is not a recent development. The ancient Romans polluted their drinking water by using lead pipes and cups. They didn't realize that lead is slightly soluble in water (see Table 9-2). The effects of lead poisoning include insanity and may have contributed to the madness of some emperors and the eventual downfall of Rome. In London, the Thames River has been polluted for hundreds of years. Strenuous efforts to clean the river has resulted in a return of some fish. Fish are returning to the Hudson River in New York City in greater numbers. Lake Erie has been considered a dead lake, unfit even for swimming. As the world population grows and technology advances, the problem of polluted water will increase. Water is polluted by the dumping of industrial wastes, oil, sewage, garbage, insecticides, detergents, and other chemicals. In 1969, an actual fire broke out on the Cuyahoga River, which flows into Lake Erie, because of excessive amounts of oil on the river water.

Almost half the population of the United States depends on water from wells. About one out of each hundred wells has some water contamination. Pollutants seep into the ground and migrate into the **aquifers** (rock through which water seeps). In overpopulated coastal areas, too much water may be drawn, and salt water may enter into the aquifer. In other cases, landfills may leak pollutants into the ground. Untreated sewage seeping into the water supply could cause the spread of disease. The costs to clean up contaminated water supplies can be very high. Aquifers feeding Denver, Colorado, are contaminated by the Rocky Mountain Arsenal. Radioactive wastes have been found in the water supply. The clean-up cost could be more than $1 billion.

Water quality can be improved by special treatment methods. Water treatment may be a combination of chemical and biological processes. Bacteria may be used to break down sewage. Ozone added to contaminated water (ozonization) at high temperature and pressure kills disease bacteria. Filtration through resin filters is often effective in removing industrial contaminants. Osmosis (diffusion through a membrane) and other separation processes will require development.

13-8b   How Do Detergents Contribute to Water Pollution?

Early detergents were not biodegradable. As a result, rivers and lakes were seen with suds on them. Sewage treatment plants were not able to operate because of detergent foam. There were no bacteria that could consume these detergents and decompose them into simpler substances. Biodegradable detergents were developed in 1965, and the foam has been slowly disappearing since that time.

Today, many detergents contain phosphates to improve their cleaning effectiveness. Unfortunately, they also serve as good fertilizers for algae (green plants that live in water). When the phosphate is dumped into a lake or river, the algae grow in large numbers. An overgrowth of algae results, and when they die, they sink to the bottom of the water. Decay organisms take large amounts of oxygen from the water as the algae decompose. As a result, there is not enough oxygen to support other life. In the

(continued on page 329)

FIGURE 13-3a

Chemical dumping at sea is no longer a viable method of toxic waste disposal.

Courtesy: EPA

FIGURE 13-3b

Where should we store this?

FIGURE 13-3c

How can we dispose of chemical wastes?

Courtesy: EPA

FIGURE 13-4a

Courtesy: U.S. Steel Corp.

Surface mining of iron ore devastates the landscape.

FIGURE 13-4b

A small sewage treatment plant in operation. This plant serves a small rural community.

FIGURE 13-4c

A large sewage treatment plant under construction in New York City.

FIGURE 13-4d

This sanitary landfill will eventually become a fertile park.

absence of oxygen, **anaerobic bacteria** (bacteria that do not require oxygen) begin to feed on the dead algae. The anaerobic decay produces foul-smelling compounds such as hydrogen sulfide (H_2S). This process, in which algae are fertilized to overgrowth and subsequently die and decay by anaerobic bacterial action, is called **eutrophication**. Eutrophy is derived from two Greek words: *eu* means "good," and *trophy* means "nutrition." Eutrophy results from the nurturing of the algae.

13-8c   How Do Metals Contribute to Water Pollution?

In 1953, many people died in Japan from mercury poisoning. These people were eating fish that had been contaminated by this metal. The mercury was being released into the water by local industries.

We have already discussed lead poisoning from automobile exhausts. Lead is also found in seawater, along with many other metals. No one knows how much of this metal a person can consume with no apparent effects. Heavy metals are difficult to eliminate from the body. While metals such as iron, copper, manganese, and zinc are needed by your body in trace amounts, lead is not desirable in any amount.

13-8d   How Is Sewage Treated?

The earth has absorbed sewage since life began. Sewage may contain disease organisms and poisonous chemicals. In the past, sewage was buried in the ground or dumped into the oceans, rivers, and lakes. The poisons were purified by the earth. Scientists seeking new antibiotics often find them in the soil itself. Today, the sewer systems of large cities are no longer able to cope with the enormous amounts of wastes they receive. For the past 60 years, New York City and communities in New Jersey have dumped sewage into the Atlantic Ocean. Today there is a dead spot 12 miles east of Sea Bright, New Jersey, that is spreading. If the dumping were to stop, it is estimated it would take ten years for the "dead spot" to cleanse itself. Expensive sewage treatment plants must be constructed. These plants accept the wastes in a primary (first) treatment stage. The sewage is aerated to provide oxygen that will enable bacteria to break down the material rapidly. In some cases, the sewage is briefly treated with chlorine to kill pathologic, or disease-causing bacteria. Fortunately, pathological organisms are more vulnerable to destruction than beneficial bacteria. As the population grows, it is becoming necessary to process sewage from a primary stage through a secondary and a tertiary (third) stage. In primary treatment, about 40% of the waste is purified. In secondary treatment, about 90% of the waste is eliminated. Tertiary treatment is the most expensive stage. Very few communities use a three-stage process, but it is becoming more necessary. The water that is produced in the third stage can be reused for irrigation and possibly drinking. The amount of sewage in water is measured in a biochemical oxygen-demand (BOD) test. A sample of sewage water is measured for its oxygen content and another sample is stored for five days at 20°C. Then, the oxygen content is measured again. The decrease in the oxygen is a measure of sewage content.

Other remedies for waste disposal are to dump the sludge further out to sea (106 miles off the coast into 8,000 feet of water). In Philadelphia, sewage is composted (decomposed for fertilizer). They call it the "Phil-organic" method. The compost is sent to strip mining areas in western Pennsylvania to help restore the landscape, or it is given to gardeners. This approach cannot be used if the sewage is contaminated with industrial wastes.

13-8e   How Do Pesticides Contribute to Water Pollution?

Pesticides are chemicals that kill insects, fungi, and weeds. During World War II, DDT was used to kill the mosquitoes that spread malaria among our soldiers. DDT is not biodegradable. It has reached into the fatty tissue of many plants and animals. This pesticide has been found in penguins at the South Pole as well as seals at the North Pole. Many bird species are threatened with extinction because of DDT. Some birds now produce eggs with extremely thin shells. The chances for the survival of the developing chick is reduced. DDT is also found in mothers' milk. An alternative approach to the control of insects is by the use of **pheromones**. These are chemicals that are excreted by female insects to attract the male. Synthetic pheromones are being made which attract undesirable male insects, enabling farmers to kill them.

13-9a   How Do Humans Pollute the Land?

The land portion of the earth is exploited and spoiled by people when topsoil is removed or destroyed, or by the disposal of solid wastes.

13-9b   How Do Humans Exploit and Destroy the Topsoil?

It is estimated that it takes 500 years for one inch of topsoil to form. The surface mining of coal and metals has changed good farm and forest land into permanent deserts. Even with reclamation, the land cannot be restored to its previous condition. The construction of concrete highways has also obliterated good land that will take hundreds of years for nature to restore.

13-9c   How Do Solid Wastes Pollute Our Land?

It is estimated that the average person produces about five pounds of solid wastes (refuse or garbage) each day. As the population increases and technology advances, the waste problem is expected to increase. There have been many attempts to recycle glass soda bottles, aluminum cans, etc., but such efforts have enjoyed limited success. Solid wastes are disposed of in two ways. They are either incinerated (burned) or buried in a **sanitary landfill**.

13-9d   How Are Solid Wastes Incinerated?

Private incinerators are a thing of the past, but they are still used in rural areas. In the cities, garbage is burned in municipal incinerators. They must meet high standards for clean air. Their major products are carbon dioxide and water. The solid fly ash is scrubbed out by electrostatic precipitators. There is a growing problem from the burning of plastics such as polyvinylchloride (PVC). The burning of chlorinated plastics results in the release of hydrogen chloride (HCl). Attention is being given to the limited use of these plastics.

13-9e   How Do Sanitary Landfills Solve the Problems of Solid Waste Disposal?

A sanitary landfill consists of refuse spread in a layer about ten feet thick. This is then covered with a thin layer of clean soil. Plastics that are not biodegradable are not suitable for burial in a landfill. The landfill is built up into a small mountain. Decay bacteria decompose the refuse. Methane and carbon dioxide are produced and seep out of the landfill into the air. These gases do not present any important problems. In some communities, efforts are being made to collect the methane for use as a fuel. Methane is the fuel used in home gas stoves (see section 13-3).

Landfills are not suitable for the construction of housing or large buildings, but when they are completed, they make excellent recreation areas. The soil is enriched from the decaying refuse and supports beautiful plants. There is a possibility that rainwater may leach the landfill and contaminate the groundwater. This has not been a problem to date.

13-9f   How Else Have People Polluted the Land?

Use of the land to bury lethal chemicals can result in an unhealthful environment. The Hooker Chemical Company buried poisonous chemicals at the Love Canal in the northwestern part of New York State. Precautions were taken that no chemicals would leak or be leached by rain into the groundwater. Several years after the burial, homes were built in the area. A school was also constructed. The construction caused a disturbance that made the chemicals leak. As a result, the school is now closed and the homes have been abandoned. The area is unfit for human habitation. The problem of cleaning up chemical dumps has yet to be solved.

13-10   How Can Poisonous Chemicals Be Eliminated?

Every year, thirty-five million tons of lethal chemical wastes are produced at 270,000 industrial sites in the United States alone. Many of these chemicals are loaded with chlorine and are not biodegradable. Some of these chemicals are polychlorinated biphenyls (PCBs), dry-cleaning fluids, pesticides, and plastics. These chemicals, if buried,

can leak into our groundwater. They can cause serious liver damage and cancer. Scientists are working intensively to develop new technologies for the elimination of these life-threatening chemicals. One hopeful idea, started in Europe, is to reduce these wastes to carbon dioxide and water by burning them in incinerator ships. This plan calls for multiple burnings at very high temperatures in ceramic ovens. This would take place 125 miles off the coast, away from all shipping lanes. The chlorine would be converted to hydrogen chloride gas. This gas dissolved in water will become hydrochloric acid. This acid seems to be absorbed into the oceans in large quantities with no apparent harm to sea life or the environment.

FIGURE 13-5

Incinerator Ship

Can we safely incinerate chemical wastes by burning them at a location in the ocean, 125 miles from the coast?

Now You Know

1. The earth is a balanced chemical environment, shared between the living and the nonliving components of the planet.

2. Humans are only one component of the chemical environment.

3. Normal air consists of nitrogen (79%), oxygen (20%), and dust (solid), water (solid, liquid, and gas), carbon dioxide, and trace elements (1%).

4. Combustion may be rapid or slow, complete or incomplete, spontaneous, or explosive.

5. Fires can be prevented and extinguished by depriving the fire of the necessary oxygen, fuel, or ignition temperature.

6. Acid rains are normal, but human pollution has increased the acid concentration of rain to more than ten times that of natural rains.

7. Carbon dioxide seems to be a harmless pollutant in our air.

8. Carbon monoxide is a human-made pollutant from automobiles. It is a poisonous gas that can cause death.

9. The automobile is a major air pollutor. It releases carbon monoxide, oxides of nitrogen, lead, and cancer-causing chemicals.

10. Water pollution is not a recent development. It was known in the days of ancient Rome.

11. Some sources of water pollution today are industrial wastes, oil, sewage, garbage, insecticides, detergents, and other chemicals.

12. Excess phosphate fertilizes algae resulting in overgrowth. When the algae die, their decay bacteria depletes the water of needed oxygen. Aquatic life is threatened by the cutoff of oxygen. Eutrophication results.

13. Detergents presented two problems. They had a high phosphate content and were not biodegradable. In 1965, chemists learned how to synthesize biodegradable detergents, but the high phosphate problem is still without a solution.

14. Poisonous metals, such as mercury and lead, that find their way into our water, eventually get into the food chain. Fish that people eat may be dangerously contaminated.

15. Raw sewage must be treated before it can be disposed of, or it may contaminate our water or food supply.

16. Many pesticides, such as DDT, are not biodegradable. They eventually get into our food chain and can poison all living things.

17. Land is polluted by the destruction of the topsoil, and by the disposal of solid wastes.

18. Topsoil is destroyed by open-pit (surface) mining operations, and by the construction of highways.

19. Solid wastes are either incinerated or buried in sanitary landfills.

20. Sanitary landfills offer the extra benefits of the development of recreational land and useable fuel gas. They present the danger that rain may leach poisonous material into the ground.

21. Another source of land pollution comes from chemical burial sites. Dangerous chemicals may seep into the groundwater, or be liberated when the land is disturbed.

New Words

aerosols	Microscopic solids and liquids suspended in the air.
anaerobic bacteria	Bacteria that carry out their life functions without oxygen.
aquifer	Water-bearing layer of rock.
biodegradable	Able to be decayed by bacteria.
BOD	Biochemical oxygen demand: a measurement of the amount of sewage in a sample.
ecology	The study of the relationship between living things and their environment.
eutrophication	The decay of dead material in the absence of oxygen.
fly ash	Unburned pieces of ash that are easily windblown.
food chain	The sequence in which organisms feed on each other.
greenhouse effect	The trapping of sunlight as heat.
pheromones	Chemicals secreted by the female insect, to attract the male.
pollution	The contamination of the environment by the introduction of a poisonous substance, or by changing the concentration of an existing substance.
saltation	The blowing of soil into the air by the wind.

Reading Power

For each of the following questions, select one answer that seems most correct.

1. The main idea of the introduction is:
   a. The earth is carrying billions of different life forms.
   b. Changes in one part of our environment cause changes in all other parts.
   c. The planet Earth is a spaceship.
   d. The earth travels 18.5 miles per second in its orbit around the sun.

2. The main idea of section 13-1 is:
   a. Normal air is not naturally polluted.
   b. Aerosols are microscopic particles in the air.
   c. Normal air is a mixture of solids, liquids, and gases.
   d. Space is dirty.

3. The main idea of section 13-2 is:

    a. Metals gain weight when they burn.
    b. Combustion, burning, and fire all mean the same thing.
    c. Paper and wood lose weight when they burn.
    d. When substances burn in air, they combine with oxygen.

4. The main idea of section 13-3 is:

    a. Combustion requires oxygen, fuel, and the necessary ignition temperature.
    b. Combustion can be rapid, slow, complete, incomplete, spontaneous, or explosive.
    c. An explosion is a very rapid and uncontrolled reaction.
    d. The rusting of iron is a slow oxidation.

5. The main idea of section 13-4 is:

    a. Fires can be prevented with care and good sense.
    b. Smothering deprives a fire of oxygen.
    c. Fires will be extinguished if they are deprived of oxygen, fuel, or the necessary ignition temperature.
    d. Never throw water on an oil fire.

6. The main idea of section 13-5a is:

    a. Rain occurs when water droplets fall to the ground.
    b. Vapor rises when water evaporates.
    c. Some rains are naturally acidic.
    d. Green plants need rain.

7. The main idea of section 13-6a is:

    a. Acid rains are natural on the earth.
    b. Green plants benefit from natural acid rain.
    c. Normal air contains dust.
    d. Air is naturally dirty and rainfall is naturally acidic.

8. The main idea of section 13-7a is:

    a. Excessive carbon dioxide produces a greenhouse effect.
    b. Burning coal produces carbon dioxide.
    c. Burning oil produces carbon dioxide.
    d. Sunlight is trapped in the air by the greenhouse effect.

9. The main idea of section 13-8a is:

    a. It is dangerous to drink water from lead containers.
    b. Water pollution is a recent problem.
    c. Water pollution has been going on for thousands of years.
    d. Fish are returning to the Hudson River.

10. The main idea of section 13-9a is:
    a. Surface mining is a major contributor to land pollution.
    b. Land is polluted by the construction of highways.
    c. It takes nature about 500 years to develop one inch of good topsoil.
    d. All of the above.

Mind Expanders

1. Why isn't water used to fight an oil fire?

2. How does fire contribute to air pollution?

3. What is the role of oxygen in burning?

4. What purpose does atmospheric dust serve?

5. How is the air cleansed by a rain or snowstorm?

6. What danger is posed by the use of lead-free fuels?

7. How is carbon monoxide dangerous to your health?

8. How are the following acid rains formed?
   a. carbonic acid   b. nitric acid   c. sulfuric acid

9. What is the greenhouse effect?

10. List three ways in which people pollute water, and at least one consequence if the practice is continued.

11. Explain how a lake can undergo eutrophication.

12. What evidence do we have that the oceans are becoming polluted?

13. How can pheromones help to control the insect population without the hazards of pollution?

14. List two ways to dispose of solid wastes. List one advantage and one disadvantage of each method.

Completion Questions

Complete each of the following statements, using the word(s) listed on the top of the next page.

ecology	carbon dioxide	lead	space	aerosols
sewage	oxygen	nitrogen	dust	carbon monoxide

1. _____ was a water pollutant in the days of ancient Rome.

2. _____ is the study of the relationship between living things and their environment.

3. The main components of our air are _____ and _____ .

4. _____ are microscopic liquid and solid particles suspended in the air.

5. _____ is the greatest source of dust in our atmosphere.

6. A major source of water pollution today is _____ .

7. A cloud is formed when droplets of water condense on _____ .

8. Human-made pollution takes place when _____ is in the air.

9. An air pollutor that probably has no harmful effects is _____ .

10. _____ is a metal pollutant released into the air by older cars.

Find the Facts

Choose the section in Chapter 13 that supports each answer to the completion questions. Copy the section number into your notebook near your answer.

Matching

Write the number in column B that relates to the word(s) in column A in the space indicated.

COLUMN A	COLUMN B
_____ pollution	1. The relationship between living things and their environment.
_____ ecology	2. Microscopic solids and liquids suspended in the air.
_____ combustion	3. Contamination of the environment.

———— greenhouse effect   4. Combining with oxygen.

———— aerosols   5. The trapping of sunlight as heat.

Multiple-Choice

1. In a balanced chemical environment:
    a. living things take chemicals from nonliving sources.
    b. living things give more than they receive from the mineral environment.
    c. living things exchange chemicals with the mineral environment equally.
    d. there is no connection between the living and the nonliving environment.

2. Normal air consists of the following components except:
    a. oxygen   b. carbon monoxide
    c. carbon dioxide   d. nitrogen

3. The following is a pollutant that is a threat to people:
    a. oxygen   b. carbon monoxide
    c. carbon dioxide   d. nitrogen

4. Most of the solid material in our atmosphere comes from:
    a. volcanoes   b. forest fires   c. coal stacks   d. outer space

5. All of the following acid rains are normal except:
    a. nitric acid   b. hydrochloric acid
    c. sulfuric acid   d. carbonic acid

6. All of the following are water pollutants except:
    a. oil   b. detergents   c. sewage   d. soap

7. The problem with detergents today is that they are:
    a. loaded with phosphates and they are not biodegradable.
    b. loaded with phosphates but they are biodegradable.
    c. not loaded with phosphates and they are not biodegradable.
    d. not loaded with phosphates and they are biodegradable.

8. All of the following metals were found in polluted water except:
    a. sodium   b. lead   c. mercury   d. gold

9. Your body needs all of the following metals in trace amounts except:
    a. manganese   b. lead   c. zinc   d. copper

10. All the following practices cause the destruction of topsoil except:
    a. strip mining
    b. highway construction
    c. sanitary landfill
    d. burial of lethal chemicals

True or False

If the statement is true, mark it *T.* If the statement is false, change the indicated word(s), using the list below, to make the statement true.

automobile	carbon dioxide
old	volcanic dust

1. Pollution of the environment is a(n) *recent* development.

2. The worst offender in air pollution is the *volcano.*

3. *Carbon dioxide* can produce a greenhouse effect and warm the earth.

4. Acid rains are *natural.*

5. The least offensive pollutant is *carbon monoxide.*

Related Science Careers		
High School Diploma	**2-Year Degree**	**4-Year or Graduate Degree**
waste water treatment plant operator	waste water treatment plant technician	waste water treatment superintendent
firefighter	fish culture technologist	oceanographer
	lab technician	arson investigator
	health inspector	insurance underwriter
	forestry technican	

Chapter 14

How Do Some Household Chemicals Work?

Instructional Objectives

After completing this chapter, you will be able to:

1. Define biodegradable, bleach, detergent, dry cleaning, emulsion, hard soap, nascent oxygen, soap, soft soap.
2. Explain how soap is made.
3. Compare different kinds of soap.
4. Explain how soap cleans.
5. Compare soap with detergent.
6. Explain how drain cleaners work.
7. Explain the advantages of ammonia as a cleanser.
8. Compare dry cleaning with wet laundering.
9. Compare bleaching by oxidation with reduction.

Chapter 14 Contents

Introduction **343**

14-1   How Is Soap Made?   **343**

14-2   What Kinds of Soaps Are There?   **344**

14-3   How Does Soap Clean?   **345**

14-4   Why Are Detergents Used in Place of Soap?   **346**

14-5   How Do Drain and Oven Cleaners Work?   **347**

14-6   Why Is Ammonia Preferable as a Cleanser?   **348**

14-7   What Is Dry Cleaning?   **348**

14-8   How Are Spots Removed from Fabrics?   **348**

14-9   How Does Bleaching Take Place?   **349**

Written Exercises   **352**

Career Information   **356**

Chapter 14

How Do Some Household Chemicals Work?

Introduction

Queen Isabella of Spain (1451–1504) claimed that she had bathed twice in her entire life—once when she was born and again on her wedding day. Queen Elizabeth I of England (1558–1603) took a bath every three months whether she "needeth it or no." Cleopatra, the beautiful queen of Egypt, bathed in fragrant oils. The oils softened her skin, and the perfumes were needed to camouflage the odors produced by the bacteria that lived on her skin. Soap was known in ancient Rome, but was too harsh to use on skin. A soap factory was revealed among the ruins of Pompeii. The use of soap and water as we know it started in London, when it was realized that poor personal hygiene was a part of the cause of cholera and typhoid epidemics. In 1846, the British government passed a Public Baths and Wash House Act. It provided public baths and laundries for the working-class people of London. The idea rapidly spread throughout Europe and the United States.

14-1 How Is Soap Made?

Soapmaking is probably the second oldest chemical process known (fermentation is the oldest). The first soap was probably made when melted fat, from a pot boiling over a wood fire, spilled onto the ashes. The contact of the hot fat with the hot ashes resulted in soap. The ashes of burned wood contain sodium and potassium oxides. These chemicals will react with water to produce sodium hydroxide (lye NaOH) and potassium hydroxide (KOH). These strong bases or alkali will react with fats to produce soap.

Table 14-1

$$Na_2O \ + \ 2 H_2O \longrightarrow 2 NaOH$$

$$K_2O \ + \ 2 H_2O \longrightarrow 2 KOH$$

Saponification is the reaction between a fat and a strong base to make soap and glycerol.

$$NaOH \ + \ Fat \longrightarrow soap \ + \ glycerol$$

FIGURE 14-1

Courtesy: Colgate-Palmolive Co.

The soap, made by reacting lye with fat, is not suitable for contact with your skin. It will dry out your skin by removing its oils. The excess lye is caustic, and will dissolve your flesh. This is probably the reason Cleopatra and the queens of Europe did not bathe with soap. Today, we wash our skin with toilet soap. The caustic (burning or dissolving) alkali is not used to excess. This soap may even have fragrances added to it. It is mild and gentle to your skin. It is a safe and effective cleansing agent.

14-2   What Kinds of Soaps Are There?

Pure soap will sink in water. Floating soaps have air frothed into them before they harden. Soap is naturally light brown in color. Toilet soaps are made white by bleaching. Many soaps have coloring matter added to them. Lava soaps contain an abrasive such as sand or ground pumice. Pumice is a light volcanic rock. Colgate's "Palmolive" soap is made from palm oil and olive oil. Soft soap contains the glycerol that is a by-product of the reaction. Soft soap is used as a shaving cream. The glycerol serves as a lubricant to allow the razor to slide easily over your skin. The soap and glycerol also serve to soften your skin. When soap is made commercially, the soap-glycerol mixture is salted out. Soap is less soluble in salt water and floats to the top. The glycerol is removed from the bottom of the container. Hard soap is the result.

Automotive grease is oil that is thickened by the addition of a lithium soap (the lithium salt of a fatty acid). It is used to lubricate those areas of a car where oil would drip or run out.

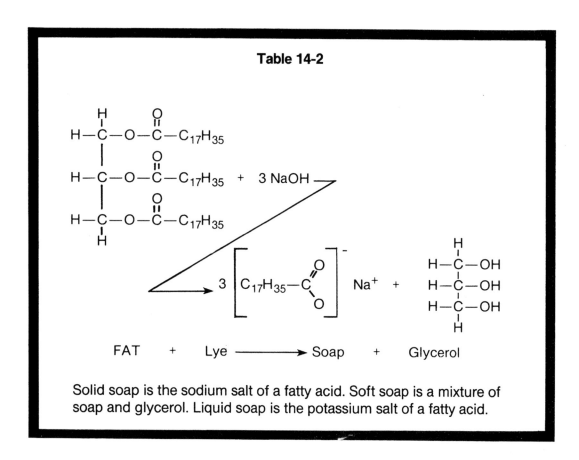

Table 14-2

FAT + Lye ⟶ Soap + Glycerol

Solid soap is the sodium salt of a fatty acid. Soft soap is a mixture of soap and glycerol. Liquid soap is the potassium salt of a fatty acid.

14-3   How Does Soap Clean?

Simple soaking of dirty clothes in water will remove some dirt from the fabric. Agitation will mechanically loosen more dirt and grime, but it will not remove all of it. Grime sticks to your skin because it adheres to your skin oils. Oil will hold dirt against water.

The molecule of soap has two parts to it. The long carbon chain is nonpolar and will dissolve in the nonpolar grease and oil (see section 5-7). The other end of the fatty acid chain is polar, and will dissolve in water. As a result, the oil is suspended in the water as small droplets. An **emulsion** (section 5-4) is formed. An emulsion is a suspension of microscopic oil droplets in water. When you wash yourself with soap, your skin oils are emulsified. The dirt that adheres to these oils are also put into suspension. The oils, along with the dirt, are flushed away under running water. Emulsifying agents are partly soluble in water, and partly soluble in oils. In general, cleansing agents are emulsifying agents.

FIGURE 14-2

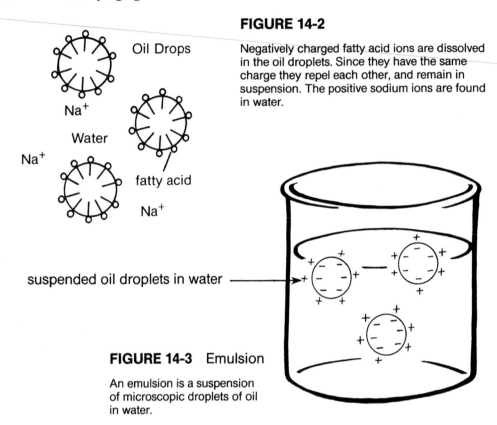

Oil Drops

Na⁺

Water

Na⁺

fatty acid

Na⁺

Negatively charged fatty acid ions are dissolved in the oil droplets. Since they have the same charge they repel each other, and remain in suspension. The positive sodium ions are found in water.

suspended oil droplets in water

FIGURE 14-3   Emulsion

An emulsion is a suspension of microscopic droplets of oil in water.

14-4   Why Are Detergents Used in Place of Soap?

All soaps will not emulsify oils in water. The calcium, magnesium, and iron salts of fatty acids are not water soluble, and will not clean clothes or skin. These salts appear as a thick jelly-like, gummy mass. In many areas, the water supply contains dissolved calcium, magnesium, or iron ions. These are **hard water** areas. Soap will not form suds with hard water. A jelly-like mass forms between the soap and the hard water minerals. It sticks to your skin, leaves a ring around your bathtub, and sticks to your washed clothes. White clothing never looks as white and colors are not as bright as they could be because of this sticky mass in the fabric. The chemist's answer to the problem is the creation of a synthetic soap, or **detergent**. Detergents wash equally as well as soap in hard or soft water. Detergents are **sulfonated** long-chain alcohols. (*Sulfonated* means that sulfuric acid has been added.)

Table 14-3

$$2 \, (C_{17}H_{35}COO^-) \, Na^+ \; + \; Ca^{++} \longrightarrow Ca(C_{17}H_{35}COO)_2\!\downarrow \; + \; 2 \, Na^+$$

SOAP + HARD WATER ⟶ GUMMY INSOLUBLE PRECIPITATE
(SOAP SCUM)

$$C_{11}H_{23}CH_2OH \; + \; H_2SO_4 \longrightarrow C_{11}H_{23}CH_2OSO_3H \; + \; H_2O$$

lauryl + sulfuric ⟶ lauryl hydrogen
alcohol   acid   sulfate

$$C_{11}H_{23}CH_2OSO_3H \; + \; NaOH \longrightarrow C_{11}H_{23}CH_2OSO^-_3Na^+ \; + \; H_2O$$

lauryl hydrogen + lye ⟶ sodium lauryl sulfate
sulfate   (detergent)

If you live in a soft-water area, you may choose to use either soap or detergent for your laundry. In hard-water areas, you must either soften your water or launder with a "soapless soap" or detergent. Early detergents created sewage problems because soil bacteria were not able to break down the detergent molecules, and thereby destroy its sudsing effect. It is important that biodegradable detergents are used. Biodegradable substances are decomposed into simpler substances by decay bacteria.

14-5 How Do Drain and Oven Cleaners Work?

All drainpipes have a U-shaped bend. The bend in the drain holds water and prevents sewer gases from entering your home. The U also prevents rats from gaining access to your home from the sewers. When grease and hair are washed down the drain, they may get caught in the U-trap. If a strong base such as sodium hydroxide is poured down the drain, the oil clog will react with the alkali to form soap. At the same time, the strong base dissolves the protein in the hair fibers. Drain cleaners may have aluminum chips added to sodium hydroxide flakes. When added to a clogged drain, the chemicals react in the water to release hydrogen gas along with much heat. The mixture sounds strong as the gas bubbles out and the drain water boils. The released gas causes agitation. Care must be taken that the hot lye solution does not spatter back and get into your eyes. The first aid for chemical burns is to flush the area with cold clean water and to call your doctor immediately.

Table 14-4

$$6\,NaOH \;+\; 2\,Al \longrightarrow 2\,Na_3AlO_3 \;+\; 3\,H_2 \;+\; heat$$

Grime buildup in ovens is also due to grease. Oven cleaners contain a strong base (lye: NaOH), enabling saponification to take place.

14-6   Why Is Ammonia Preferable as a Cleanser?

Ammonia is a weak base and less dangerous to use than lye (NaOH). It is a gas that is extremely soluble in water. Ammonia solutions leave no residue because they completely evaporate. Ammonia is an emulsifying agent and a good cleanser for glass (windows and mirrors) as well as kitchen linoleums. It should be used in well-ventilated areas to allow the fumes to escape.

14-7   What Is Dry Cleaning?

Dry cleaning is the removal of dirt without any water. It is used where water would cause staining or other damage, as with silk. The cleansing is done when the grease holding the dirt is removed by oil solvents. There are many dry-cleaning fluids such as ether, naphtha, benzine, gasoline, carbon tetrachloride, dichloroethylene, etc. All of these fluids are dangerous because they are explosive when mixed with air, and very combustible, with the exception of carbon tetrachloride. Carbon tetrachloride is dangerous because it is carcinogenic (causes cancer).

14-8   How Are Spots Removed from Fabrics?

The rules for spot removal are as follows:

1.   Act as soon as possible. Fresh stains are easier to remove than old stains.

2.   Determine the cause of the stain. All stains are not treated in the same way.

3.   Determine the fabric that is stained. All fabrics cannot be treated the same way.

4.   Do not get complicated. Simple methods usually get the best results.

5.   Be careful. If you are not sure of the stain or the fabric, make a test on a small sample first. You can make matters worse.

Table 14-5   Spot and Stain

Stain	Treatment
acids	Cover with sodium bicarbonate. Rinse with warm water.
ballpoint ink	Soak in rubbing alcohol before laundering.
blood	Sponge with cold water as soon as possible. Soak in an enzyme presoak solution and launder.
candy	Soak in laundry soap and chlorine bleach. Launder as usual.
grass	Sponge with alcohol. Launder as usual.
gum	Apply an ice cube. Chip off the brittle gum.
ink-stained fingers	Rub with the sulfur end of a match.
wax	Same as gum or launder in hot water.

14-9   How Does Bleaching Take Place?

Bleaching is the removal of color from a material. Some materials, such as hair, can be bleached by exposure to the sun. Most materials are bleached with chemicals. Bleaching takes place by **oxidation** or by **reduction**.

Chlorine and its compounds bleach by oxidation. Laundry bleaches contain a 5% sodium hypochlorite (NaClO) solution in water. A 16% solution is sold for use in swimming pools, and is called **swimming pool chlorine**. It is too concentrated for household use. Sodium hypochlorite (Javelle water) is the liquid laundry bleach sold in supermarkets. It can be made by passing an electric current through a common saltwater solution.

Table 14-6

$$NaCl + H_2O \xrightarrow{\text{ELECTRICITY}} NaClO + H_2$$

salt + water —— ELECTRICITY ——▶ sodium hypochlorite + hydrogen

Sodium hypochlorite solutions are unstable and will break down with time. It is preferable to purchase this bleach as a powder because it can stand longer without losing its strength. Ordinary bleaching powder is calcium hypochlorite: $Ca(ClO)_2$ or $CaCl(ClO)$. It is also known as **chloride of lime**.

Chlorine becomes a very strong bleach in the presence of water. The hypochlorous acid formed is unstable and releases **nascent** (atomic) **oxygen**. It is this atomic oxygen that is responsible for the bleaching action of chlorine.

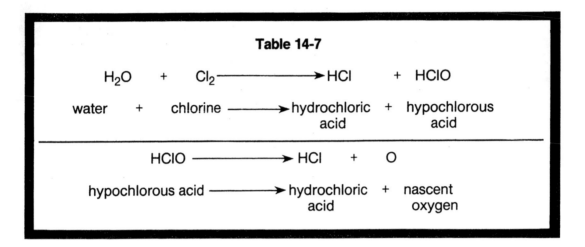

Table 14-7

$$H_2O \ + \ Cl_2 \longrightarrow HCl \ + \ HClO$$

water + chlorine ⟶ hydrochloric + hypochlorous
acid   acid

$$HClO \longrightarrow HCl \ + \ O$$

hypochlorous acid ⟶ hydrochloric + nascent
acid   oxygen

The word nascent comes from Latin and means "newborn." Nascent (atomic) oxygen is very reactive. Chlorine bleaches are very harsh and shorten the life of the fabric.

Delicate materials such as straw, silks, paper, and dried fruits are bleached by reduction. The reducing agent is sulfur dioxide (SO_2), or its water solution, sulfurous acid (H_2SO_3). Materials that are bleached by reduction will usually darken when exposed to air and sunlight over a long period of time. The oxygen in the air slowly oxidizes the material, thereby undoing the effects of reduction. This explains how white paper yellows with age, and straw hats darken.

Hydrogen peroxide and sulfur dioxide bleaches are considered mild. Chlorine bleaches tend to destroy or rot the fibers of the bleached garment and are destructive to living tissue. Chlorine is poisonous and one should follow the directions listed on the label.

Table 14-8

$$SO_2 \ + \ H_2O \longrightarrow H_2SO_3$$

sulfur dioxide + water ⟶ sulfurous acid

Sulfur dioxide bleaches by reduction.

$$2\,H_2O_2 \longrightarrow 2\,H_2O \ + \ O_2$$

hydrogen peroxide ⟶ water + oxygen

Hydrogen peroxide decomposes to release oxygen. It is used to bleach hair.

Now You Know

1. Soap has been made for thousands of years, but the early soaps were too harsh to be used on human skin.

2. Saponification is the reaction between a fat and a strong base to yield soap and glycerol.

$$\text{fat} \ + \ \text{lye} \longrightarrow \text{soap} \ + \ \text{glycerol}$$

3. Soft soap is a mixture of soap with glycerol.

4. An emulsion is a suspension of droplets of oil in water.

5. Soap cleans by emulsifying oil and grease, along with the dirt that sticks to it. The emulsion is then flushed away under running water.

6. Detergents are synthetic, soapless soaps. They function well in hard water.

7. Soaps are not able to make suds, or emulsify oils, in hard water.

8. Hard water contains calcium, magnesium, or iron ions.

9. Drain cleaners and oven cleaners are composed of strong bases. They change grease and oil clogs into soap. Strong bases also dissolve hair.

10. Ammonia is popular as a cleansing agent because it is less dangerous than harsh bases and it leaves no residue.

11. Dry cleaning means cleaning without water. Oil solvents are used. These solvents can be dangerous to your health.

12. Spots and stains are removed in specific ways. Care must be used or the garment may be ruined.

13. Bleaching is the removal of color from a material. Some dyes are made colorless by oxidation. Other dyes can be made colorless by reduction.

New Words

biodegradable	Capable of being broken down by decay bacteria.
bleaching	The removal of color from a material.
detergent	A synthetic soap. It forms suds in hard water.
dry cleaning	Removal of dirt without water. Oil solvents are used to remove oils, greases, and dirt.
emulsion	A suspension of oil in water.
hard soap	Soap without glycerol.
nascent oxygen	"Newborn," or atomic, oxygen. Nascent oxygen is unstable and very reactive.
soap	The metallic salt of a fatty acid.
soft soap	A mixture of soap and glycerol.

Reading Power

For each of the following questions, select one answer that seems most correct.

1. The main idea of the introduction is:
   a. Fashionable women used perfume to cover the odors of their bodies.
   b. Soap was available for more than 2,000 years, but it was too harsh for use on human skin.
   c. Soap is a recent development.
   d. Bacteria live on your skin.

2. The main idea of section 14-1 is:

   a. Lye (NaOH) can be derived from the ashes of burned wood.
   b. Glycerol is a by-product of soapmaking.
   c. Soap is the product of a fat with a strong base.
   d. Soap was probably first made from the reaction of boiled fat on the ashes of a wood fire.

3. The main idea of section 14-2 is:

   a. There is more than one kind of soap.
   b. Toilet soaps are suitable for use on our skin.
   c. Automotive grease is oil thickened with a lithium soap.
   d. Pure soap will sink in water.

4. The main idea of section 14-3 is:

   a. Soft soap is a mixture of soap and glycerol.
   b. Soap is an emulsifying agent.
   c. The soap molecule has a fat-soluble part, and a water-soluble part.
   d. Soap cleans by emulsifying oily dirt.

5. The main idea of section 14-4 is:

   a. Soap cannot function efficiently in hard water.
   b. Hard water contains calcium, magnesium, or iron ions.
   c. Detergents are soapless soaps that function efficiently in hard water.
   d. Calcium salts of fatty acids do not emulsify oils.

6. The main idea of section 14-5 is:

   a. Drain cleaners are composed of strong bases.
   b. Hair, and other proteins, are dissolved by lye (NaOH).
   c. Oven cleaners can only change grease into soap.
   d. Drain and oven cleaners saponify grease and dissolve proteins.

7. The main idea of section 14-6 is:

   a. Ammonia is a weak base and safer to use than lye.
   b. Ammonia is extremely soluble in water.
   c. Ammonia is a good emulsifying agent. It leaves no residue, and it is safer than lye.
   d. Ammonia should be used in well-ventilated rooms.

8. The main idea of section 14-7 is:

   a. Dry cleaning is the removal of dirt without water.
   b. Fluids used in dry cleaning are hazardous to your health.
   c. Some fabrics must only be dry cleaned.
   d. Soap and water clean more effectively than dry cleaning.

9. The main idea of section 14-8 is:

    a. Fresh stains are easier to remove than old stains.
    b. There are five rules for spot removal.
    c. Grass stains are usually removed with the application of alcohol or soap.
    d. Blood stains should be washed with cold water while the stain is fresh.

10. The main idea of section 14-9 is:

    a. Bleaching can take place either by oxidation or reduction.
    b. Chlorine is the strongest bleaching agent used.
    c. Delicate fabrics are bleached by reduction with sulfurous acid (H_2SO_3).
    d. Chlorine bleaches are poisonous.

Mind Expanders

1. How did the pioneers, who settled the West, get their soap?

2. Why couldn't people bathe with soap and water in ancient Egypt?

3. Explain how soap cleans.

4. How can lye, as a drain cleaner, cause pipe joints to leak? (*Hint:* Pipe joints are sealed with grease.)

5. Why were detergents developed?

6. Why should you use biodegradable detergents?

Completion Questions

Complete each of the following statements, using the word(s) below.

dry cleaning	oxidation	soap
lye	detergent	

1. _____ is the sodium salt of a fatty acid.

2. Bleaching takes place by _____ or by reduction.

3. _____ is the process of removing dirt in the absence of water.

4. _____ is suitable for use as a drain cleaner.

5. _____ is a soap substitute that is effective in hard water.

Multiple-Choice

1. All of the following soaps are suitable for washing except:

   a. sodium stearate
   b. potassium stearate
   c. calcium stearate
   d. lithium stearate

2. Hard water:

   a. does not form suds with soap
   b. is useful as a bleach
   c. does not form suds with detergents
   d. is useful in dry cleaning

3. The best way to remove dirt from clothing is by:

   a. soaking in water
   b. soaking and agitation in water
   c. emulsification by a soap
   d. agitation alone to shake dirt loose

4. All of the following are good emulsifying agents except:

   a. ammonia   b. soap   c. detergents   d. water

5. The following are all bleaching agents except:

   a. sulfurous acid: H_2SO_3
   b. hydrogen peroxide: H_2O_2
   c. sodium hypochlorite: $NaClO$
   d. sodium chloride: $NaCl$

True or False

Mark each statement true (*T*) or false (*F*). If the statement is false, replace the indicated word(s), using the list below, to make the statement true.

soap	fragrant oils	sulfur dioxide

1. In the past, clean people bathed with *soap* and water.

2. Silk, paper, dried fruits, and straw are all bleached by *chlorine.*

3. *Soap* can be made by mixing hot fat with the ashes from burned wood.

4. A by-product in the manufacture of *detergents* is glycerol.

5. Glycerol mixed with *soap* makes a good shaving cream.

Find the Facts

Choose the section in Chapter 14 that supports each answer to the completion statements, multiple-choice statements, and the true or false statements. Copy the section number into your notebook near your answer.

Matching

Write the number in column B that relates to the word(s) in column A in the space indicated.

COLUMN A	COLUMN B
_____ soap	1. "Newborn," or atomic, oxygen.
_____ emulsion	2. Capable of being broken down by decay bacteria.
_____ detergent	3. The metallic salt of a fatty acid.
_____ biodegradable	4. A synthetic soapless soap that is effective in hard water.
_____ dry cleaning	5. A suspension of oil in water.
_____ bleaching	6. Removal of dirt in the absence of water.
_____ nascent oxygen	7. The removal of color by oxidation or reduction.

Career Opportunities

High School Diploma	2-Year Degree	4-Year or Graduate Degree
household worker	laboratory technician	chemist
porter		environmental engineer
janitor	custodian	custodial engineer

Answer Section

Chapter 1 Exercises

Metric Math Problems

1. 750 milliliters of water weigh 750 grams.

2. 55 grams of water occupy 55 milliliters.

3. The resulting solution would weigh 200 grams, and occupy about 200 milliliters.

4. 150 grams of water are needed to fill the box.

5. The water-filled box would weigh 250 grams.

6. The volume of the box is 1,200 milliliters.

7. The box is 120 centimeters high.

8. The thermos can hold 700 milliliters of liquid.

9. The teaspoon can hold 5 grams of water.

10. 200 teaspoons of water are needed to fill a liter flask.

Reading Power

1.	a	3.	d	5.	c	7.	d
2.	c	4.	c	6.	a		

Mind Expanders

Answers will vary.

Sequence of Events   2, 1, 5, 4, 3   (N.B., steps 1 and 4 are interchangeable)

True or False

Intro.	1.	scientist	Intro.	6.	T	
1-1	2.	T	1-1	7.	theory	
1-2	3.	make observations and tests	1-5	8.	metric system	
1-6	4.	equally as	1-7	9.	T	
1-7	5.	T	1-1	10.	T	

Find the Facts

(The answers are listed with the True/False answers above.)

Multiple-Choice

1. d	4. c	7. c	10. d
2. d	5. d	8. d	
3. b	6. a	9. b	

Lab Safety Test

1. false	6. never	11. false	16. false
2. T	7. T	12. false	17. false
3. T	8. T	13. T	18. T
4. may not	9. never	14. T	19. always
5. T	10. T	15. T	20. always

Metric-English Conversion Problems

2 inches = 5.08 centimeters	20 kilograms = 44 pounds
226.8 grams = ½ pound	10 pounds = 4.536 kilograms
10 liters = 10.6 quarts	10 ounces = 283.5 grams
10 kilometers = 6.2 miles	4 gallons = 15.16 liters

Chapter 2 Exercises

Reading Power

1. a	3. c	5. d	7. a
2. d	4. d	6. c	

Mind Expanders

Answers will vary.

Physical and Chemical Changes

1. Breaking a vase is a *physical change.*
2. Freezing water is a *physical change.*
3. Burning paper is a *chemical change.*
4. Tearing paper is a *physical change.*
5. Dissolving instant tea in water is a *physical change.*

Completion Questions

1. matter, energy
2. matter
3. properties
4. solid, liquid, gas
5. occupies a definite space; has a definite shape
6. occupies a definite space; has no shape
7. does not occupy a definite space; has no shape
8. physical; chemical

9. physical

10. physical

11. evaporation

12. kinetic molecular theory

13. temperature

14. molecules

15. melting

16. molecules

17. freezing

18. condensation

19. condensation

20 sublimation

Matching

3. kinetic molecular theory

1. sublimation

5. freezing

4. melting

2. temperature

Chapter 3 Exercises

Reading Power

1. a

2. b

3. d

4. c

5. b

6. b

7. a

8. c

9. a

10. a

11. b

Mind Expanders

Answers will vary.

Complete the Following Table

Table 3-8

Element	Symbol	Element	Symbol
hydrogen	H	sodium	Na
lithium	Li	aluminum	Al
boron	B	phosphorous	P
nitrogen	N	chlorine	Cl
fluorine	F	potassium	K
iron	Fe	gold	Au
copper	Cu	silver	Ag

Completion Questions

1. element
2. different proton numbers
3. molecules
4. John J. Thompson
5. electron
6. protons
7. neutron
8. proton, neutron
9. electron
10. proton

Drawing Atomic Diagrams

Table 3-9

$_1H^1$

Atomic Weight ⟶ 4
Atomic Number ⟶ 2 He

2+ / 2n) 2

$_3Li^7$	$_4Be^9$	$_5B^{11}$	$_6C^{12}$	$_7N^{14}$	$_8O^{16}$	$_9F^{19}$	$_{10}Ne^{20}$
3+ 4n) 2) 1	4+ 5n) 2) 2	5+ 6n) 2) 3	6+ 6n) 2) 4	7+ 7n) 2) 5	8+ 8n) 2) 6	9+ 10n) 2) 7	10+ 10n) 2) 8

$_{11}Na^{23}$	$_{12}Mg^{24}$	$_{13}Al^{27}$	$_{14}Si^{28}$	$_{15}P^{31}$	$_{16}S^{32}$	$_{17}Cl^{35}$	$_{18}Ar^{40}$
11+ 12n) 2) 8) 1	12+ 12n) 2) 8) 2	13+ 14n) 2) 8) 3	14+ 14n) 2) 8) 4	15+ 16n) 2) 8) 5	16+ 16n) 2) 8) 6	17+ 18n) 2) 8) 7	18+ 22n) 2) 8) 8

$_{19}K^{39}$	$_{20}Ca^{40}$
19+ 20n) 2) 8) 8) 1	20+ 20n) 2) 8) 8) 2

Multiple-Choice

1. c
2. a
3. b
4. d
5. b
6. d
7. a
8. b
9. c
10. b
11. b
12. d
13. d
14. c
15. d

Chapter 4 Exercises

Part I

Electron-Dot Diagrams

1. hydrogen sulfide H_2S H:S̈:
                                       H

2. methyl chloride CH_3Cl H
                                                    H:C̈:C̈l:
                                                       H

3. hydrogen iodide   HI   H:Ï:

4. ammonia NH_3 H:N̈:H
                                                H

5. methane (natural gas) CH_4 H
                                        H:C̈:H
                                           H

Formulas

Formula	Element	# Atoms
CO_2	carbon	1
	oxygen	2
$NaNO_3$	sodium	1
	nitrogen	1
	oxygen	3
$AlPO_4$	aluminum	1
	phosphorous	1
	oxygen	4
H_2SO_4	hydrogen	2
	sulfur	1
	oxygen	4
K_2CO_3	potassium	2
	carbon	1
	oxygen	3
$MgCl_2$	magnesium	1
	chlorine	2
SiO_2	silicon	1
	oxygen	2
LiF	lithium	1
	fluorine	1
CaO	calcium	1
	oxygen	1

Formula Writing Drill

	OH^{-1}	SO_4^{-2}	PO_4^{-3}	S^{-2}	Cl^{-1}
Na^{+1}	NaOH	Na_2SO_4	Na_3PO_4	Na_2S	NaCl
Mg^{+2}	$Mg(OH)_2$	$MgSO_4$	$Mg_3(PO_4)_2$	MgS	$MgCl_2$
Al^{+3}	$Al(OH)_3$	$Al_2(SO_4)_3$	$AlPO_4$	Al_2S_3	$AlCl_3$
$(NH_4)^+$	NH_4OH	$(NH_4)_2SO_4$	$(NH_4)_3PO_4$	$(NH_4)_2S$	NH_4Cl

Reading Power

1. c
2. a. sentence 2
   b. sentence 3
3. d
4. a. sentence 4
   b. sentence 8
   c. sentence 9

5. c
6. a. sentence 9
   b. table salt (NaCl)
   c. polar bond
7. c
8. sentence 5

9. b
10. a. section 4-5
    b. section 4-6
    c. section 4-4
    d. section 4-6
11. d

12. d
13. b
14. a
15. a

Mind Expanders

Answers will vary.

Formulas and Valences of Radicals

Radical	Formula	Valence
phosphate	PO_4	–3
nitrate	NO_3	–1
carbonate	CO_3	–2
sulfate	SO_4	–2
bisulfate	HSO_4	–1
hydroxide	OH	–1
cyanide	CN	–1
acetate	$C_2H_3O_2$	–1
ammonium	NH_4	+1
bicarbonate	HCO_3	–1

Completion Questions

1. valence; atomic number; symbol
2. different
3. molecule
4. attract
5. valence

6. valence
7. lose
8. ions
9. polar bond
10. oppositely

11. symbol
12. subscript
13. valence
14. metals
15. nonmetals

Multiple-Choice

1. a
2. a
3. b
4. d
5. c

Matching

4. nonpolar covalent bond
3. nonpolar

2. covalent
1. electron-dot symbol

Complete the Following Table

Name	Formula	Number of Atoms	Number of Ions
ferric chloride	$FeCl_3$	4	4
sodium phosphate	Na_3PO_4	8	4
calcium bicarbonate	$Ca(HCO_3)_2$	11	3
ammonium sulfate	$(NH_4)_2SO_4$	15	3
magnesium hydroxide	$Mg(OH)_2$	5	3

True or False

1. T
2. different
3. T
4. metals

5. T
6. electron sea
7. T
8. nonpolar covalent bonds

9. polar covalent bond
10. metals

Part II

Formula Weights

Formula	Formula Weight	Formula	Formula Weight
HCl	36.5	$FeCl_2$	127
KBr	119	AlI_3	408
CaO	56	$Al(OH)_3$	78
NaOH	40	$Ba_3(PO_4)_2$	601
$CaCl_2$	110	$Fe_2(CO_3)_3$	292
ZnF_2	103		

Compounds

Formula	%	Formula	%
CO_2	27.3% carbon 72.7% oxygen	$CuSO_4$	40.0% copper 20.0% sulfur 40.0% oxygen
$KClO_3$	32.0% potassium 28.7% chlorine 39.3% oxygen	$Zn(NO_3)_2$	34.4% zinc 14.8% nitrogen 50.8% oxygen
$CaCO_3$	40.0% calcium 12.0% carbon 48.0% oxygen	Fe_2O_3	70.0% iron 30.0% oxygen
Na_2CO_3	43.4% sodium 11.3% carbon 45.3% oxygen	$AlPO_4$	22.1% aluminum 25.4% phosphorous 52.5% oxygen
K_2CrO_4	40.2% potassium 26.8% chromium 33.0% oxygen	$CuCl_2$	47.4% copper 52.6% chlorine
NH_3	82.4% nitrogen 17.6% hydrogen	$Ba(OH)_2$	80.1% barium 18.7% oxygen 1.2% hydrogen

Empirical Formulas

1. FeO
2. NaCl
3. $CaCl_2$
4. P_2O_5
5. NO_2

6. CH_2
7. Al_2S_3
8. NH_3
9. CH_2O
10. Na_3PO_4

11. K_2CrO_4
12. $NaMnO_4$
13. $CaSiO_3$
14. H_2SO_4
15. HNO_3

Counting Molecules

1. 1,204,000 billion billion molecules
2. 301,000 billion billion molecules
3. 903,000 billion billion molecules

4. 602,000 billion billion molecules
5. 301,000 billion billion molecules

Scientific Notation

1. 1.0×10^{-22} grams
2. 1.08×10^{22} atoms per gram
3. 6.0×10^{26} grams
4. 9.0×10^{-33} grams
5. 3.0×10^8 meters per second

6. 9.3×10^7 miles
7. 5.0×10^{-9} centimeters
8. 1.8×10^{-7} molecules
9. 4.5×10^9 people
10. 1.0×10^{-7} molecules

Chapter 5 Exercises

Part I

Reading Power

1. a
2. b
3. d
4. a

5. d
6. c
7. a
8. b

9. d
10. a
11. a
12. d

13. c
14. d
15. b

Mind Expanders

Answers will vary.

True or False

1. seawater
2. T
3. T
4. fresh water

5. T
6. solvent
7. dilute
8. solids

9. T
10. decreased

Matching

7.	tincture	2.	residue
6.	symmetric	10.	solvent
1.	suspension	9.	distillate
5.	unsaturated	4.	filtrate
8.	solute	3.	colloid

Multiple-Choice

1.	c	6.	a	11.	c	16.	d
2.	d	7.	c	12.	a	17.	c
3.	c	8.	d	13.	b	18.	a
4.	b	9.	a	14.	d	19.	c
5.	a	10.	d	15.	b	20.	b

Graph Interpretation

1.	$56°C$	6.	50 grams
2.	$Ce_2(SO_4)_3$	7.	10 grams
3.	KNO_3	8.	KI
4.	NaCl	9.	$KClO_3$
5.	130 grams	10.	700 grams

Completion Questions

1.	seawater	6.	suspensions
2.	solute, solvent	7.	saturated
3.	solute	8.	increase
4.	alloys	9.	increase
5.	distillation	10.	miscible

Part II

Molarity

Substance	Molecular Weight	Weight of Substance (grams)	Number of moles	Volume (ml)	Molarity (M)
NaCl	58	174	3	500	6
$C_6H_{12}O_6$	180	90	½	1,000	½
$CaCl_2$	111	333	3	250	12
H_2SO_4	98	196	2	2,000	1
K_2CO_3	138	552	4	1,500	2.67
NH_4Cl	53	53	1	200	5
$Al(NO_3)_3$	213	426	2	100	20
NaOH	40	120	3	50	60
$NaNO_3$	85	170	2	250	8
HCl	36	144	4	500	8

Chapter 6 Exercises

Reading Power

1. a	3. c	5. c	7. d
2. d	4. b	6. b	

Mind Expanders

Answers will vary.

Labeling (classifying) Acids, Bases, and Salts

b. NaOH	d. $Mg(OH)_2$	e. $Al_2(SO_4)_3$
a. HNO_3	g. $LiNO_3$	e. $ZnSO_4$
c. H_3PO_4	c. $HC_2H_3O_2$	
f. K_2CO_3	f. $NaC_2H_3O_2$	

Table 6-12			
Solution	Acidic, Alkaline or Neutral	Effect on Litmus	pH (more, less or equal to 7)
Na_3PO_4	alkaline	turns blue	> 7
H_2O	neutral	none	$= 7$
LiOH	alkaline	turns blue	> 7
HI	acidic	turns red	< 7
$CuSO_4$	acidic	turns red	< 7
$Pb(NO_3)_2$	acidic	turns red	< 7

Completion Questions (Neutralization Reactions)

1. $3\,NaOH$ $+$ $H_3PO_4 \longrightarrow Na_3PO_4$ $+$ $3\,H_2O$
    strong   weak   alkaline
    base   acid   salt
    sodium   phosphoric   sodium   water
    hydroxide   acid   phosphate

2. $Fe(OH)_3$ $+$ $3\,HCl \longrightarrow FeCl_3$ $+$ $3\,H_2O$
    weak   strong   acidic
    base   acid   salt
    iron (III)   hydrochloric   iron (III)   water
    hydroxide   acid   chloride

3. $2\,KOH$ $+$ $H_2SO_4 \longrightarrow K_2SO_4$ $+$ H_2O
    strong   strong   neutral
    base   acid   salt
    potassium   sulfuric   potassium   water
    hydroxide   acid   sulfate

4. KOH $+$ $HNO_3 \longrightarrow KNO_3$ $+$ H_2O
    strong   strong   neutral
    base   acid   salt
    potassium   nitric   potassium   water
    hydroxide   acid   nitrate

5. $NaOH$ $+$ $HCl \longrightarrow NaCl$ $+$ H_2O
    strong   strong   neutral
    base   acid   salt
    sodium   hydrochloric   sodium   water
    hydroxide   acid   chloride

Multiple-Choice

1. c   2. a   3. b   4. a   5. d

Table 6-15

pH	$(H)^+$ Molarity	pOH	$(OH)^-$ Molarity	Acidic Basic Neutral
3	10^{-3}	11	10^{-11}	acidic
1	10^{-1}	13	10^{-13}	acidic
12	10^{-12}	2	10^{-2}	basic
2	10^{-2}	12	10^{-12}	acidic
4	10^{-4}	10	10^{-10}	acidic
13	10^{-13}	1	10^{-1}	basic
9	10^{-9}	5	10^{-5}	basic
11	10^{-11}	3	10^{-3}	basic
7	10^{-7}	7	10^{-7}	neutral
5	10^{-5}	9	10^{-9}	acidic
10	10^{-10}	4	10^{-4}	basic
8	10^{-8}	6	10^{-6}	basic
6	10^{-6}	8	10^{-8}	acidic

Chapter 7 Exercises

Reading Power

1. d   5. a   9. d   13. a
2. c   6. d   10. a   14. b
3. a   7. c   11. c   15. c
4. b   8. b   12. d   16. b

Mind Expanders

Answers will vary.

Completion Questions

Intro.	1.	energy		7-1	6.	fossil fuels
7-1	2.	oil, coal, food		7-2	7.	nuclear fusion
7-1	3.	the sun		7-15	8.	nuclear fusion
Table 7-1	4.	food		7-3	9.	coal
Intro.	5.	oil				

Find the Facts

(The sections are indicated next to each answer above.)

True or False

1.	energy	4.	T	7.	T	10.	carbon fuels or fossil fuels
2.	oil	5.	T	8.	T		
3.	the sun	6.	green plants	9.	fossil fuels		

Matching

2.	solar energy		5.	nuclear reactor
4.	geothermal energy		8.	energy
6.	atomic fission		1.	fossil fuel
3.	fuel		7.	atomic fusion

Multiple-Choice

1.	d	4.	b	7.	c
2.	b	5.	c	8.	d
3.	a	6.	c	9.	c

Half-Life Problems

1. 3 milligrams
2. 43 days
3. The papyrus is genuine.
   It is about 5,730 years old.
4. 1.5 milligrams
5. 1 milligram

Transmutation Equations

1.	$_2He^4$	3.	$_{39}Y^{90}$	5.	$_{46}Pd^{107}$
2.	$_{-1}e^0$	4.	$_{86}Rn^{222}$		

Chapter 8 Exercises

Reading Power

1. c   3. a   5. d   7. b
2. a   4. b   6. c

Mind Expanders

Answers will vary.

Completion Questions

8-2 1. cement   8-6 4. plaster of paris
8-1 2. oxygen, silicon, aluminum 8-2 5. concrete
8-3 3. glaze

Find the Facts

(The sections are indicated next to each answer above.)

True or False

1. T   2. steel rods   3. feldspar   4. T   5. glass

Matching

5. mineral   4. Portland cement
7. silica   2. mortar
10. rock   3. plaster of paris
6. clay   8. limestone
1. cement   9. slag

Multiple-Choice

1. a   2. b   3. c   4. d   5. d

Chapter 9 Exercises

Part I

Reading Power

1. d   5. a   8. c   11. a
2. a   6. c   9. a   12. a
3. c   7. b   10. d   13. b
4. d

Mind Expanders

Answers will vary.

Completion Questions

9-1 1. ore

9-7 2. tempering

9-5 3. cast iron

9-4 4. flotation

9-4 5. taconite

Find the Facts

(The sections are indicated next to each answer above.)

True or False

1. T
2. metals
3. T
4. flotation
5. pig iron
6. flux
7. T
8. steel
9. T
10. tempering

Multiple-Choice

1. d
2. c
3. b
4. c
5. c

Matching

3. ore of iron
5. ore of copper
6. ore of uranium
1. ore of aluminum
2. ore of lead
7. ore of titanium
4. ore of zinc

Part II

Reading Power

1. c
2. c
3. a
4. b
5. a
6. a
7. d

Mind Expanders

Answers will vary.

Completion Questions

9-9 1. roasted

9-10 2. calcined

9-15 3. titanium

9-10 4. bronze

9-12 5. bauxite

Find the Facts

(The sections are indicated next to each answer above.)

True or False

1. reverberatory furnace
2. T
3. aluminum

4. titanium
5. T

Multiple-Choice

1. c   2. a   3. d   4. c   5 b

Matching

5. roasting
3. calcine
4. blister copper

2. alumina
1. reverberatory furnace

Chapter 10 Exercises

Part I

Formula Determination Problems

Alkane Formulas

$C_{21}H_{44}$ $C_{33}H_{68}$ $C_{55}H_{112}$
$C_{38}H_{78}$ $C_{32}H_{66}$ $C_{26}H_{54}$

Alkene Formulas

$C_{21}H_{42}$ $C_{33}H_{66}$ $C_{55}H_{110}$
$C_{39}H_{78}$ $C_{33}H_{66}$ $C_{27}H_{54}$

Alkyne Formulas

$C_{21}H_{40}$ $C_{33}H_{64}$ $C_{55}H_{108}$
$C_{40}H_{78}$ $C_{34}H_{66}$ $C_{28}H_{54}$

Reading Power

1. d   4. d   7. d
2. c   5. b   8. a
3. d   6. a   9. b

Mind Expanders

Answers will vary.

Matching

4. $C_nH_{(2n + 2)}$
2. C_5H_{10}
6. C_6H_6

3. C_5H_{12}
1. C_5H_8
5. Aliphatic

Table 10-12

Name	Structural Formula
1-butene	$\underset{\displaystyle\overset{\mid}{H}}{\overset{\displaystyle\overset{H}{\mid}}{C}} = \underset{}{\overset{\displaystyle\overset{H}{\mid}}{C}} - \underset{\displaystyle\overset{\mid}{H}}{\overset{\displaystyle\overset{H}{\mid}}{C}} - \underset{\displaystyle\overset{\mid}{H}}{\overset{\displaystyle\overset{H}{\mid}}{C}} - H$
isobutane	$H - \overset{\displaystyle\overset{H}{\mid}}{\underset{\displaystyle\overset{\mid}{H}}{C}} - \overset{\displaystyle\overset{H}{\mid}}{\underset{HCH}{C}} - \overset{\displaystyle\overset{H}{\mid}}{\underset{\displaystyle\overset{\mid}{H}}{C}} - H$
1-propyne	$HC \equiv C - CH_3$
pentane	$H_3C - CH_2 - CH_2 - CH_2 - CH_3$
3-octene	$H_3C - CH_2 - CH = CH - CH_2 - CH_2 - CH_2 - CH_3$

Multiple-Choice

Intro.	1.	b		10-3	6.	a
10-3	2.	c		10-5	7.	c
10-4	3.	c		10-8	8.	d
10-7	4.	d		10-4	9.	c
10-10	5.	a		10-4	10.	c

Find the Facts

(The sections are indicated next to each answer above.)

Completion Questions

1. C_4H_6 3. C_3H_8

2. C_3H_8 4. CH_2

True or False

1. alkene series   3. addition   5. biochemistry

2. T   4. T

Part II

Reading Power

1. b   2. d   3. a   4. b   5. c

Multiple-Choice

10-14   1. c   10-13   3. d   10-5   5. a
10-10   2. b   10-14   4. c

Mind Expanders

Answers will vary.

Find the Facts

(The sections are indicated next to each answer above.)

Completion Questions

1. fermentation   5. alcohol   9. carboxyl group
2. denatured   6. reduction   10. —CHO
3. reduction   7. ester
4. organic radicals   8. functional group

True or False

1. T   3. oxidation   5. ethanol
2. carboxyl group (—COOH)   4. T

Table 10-14		
Type of Compound	**Functional Group**	**General Formula**
hydrocarbon	none	R—H
alcohol	—OH	R—OH
aldehyde	—CHO	R—CHO
ketone	—CO—	R—CO—R′
acid	—COOH	R—COOH
ester	—COOR′	R—COOR′

Chapter 11 Exercises

Reading Power

1. d	6. d	11. b	16. d
2. c	7. b	12. a	17. c
3. a	8. a	13. a	18. b
4. c	9. c	14. a	19. b
5. a	10. b	15. d	20. d

Complete the Table

Vitamin	Importance	Food Sources	Miscellaneous Information
C	prevents scurvy	citrus fruits	water soluble heat sensitive
A	prevents night blindness	liver, eggs	fat soluble
B-1 (thiamine)	nervous system	whole-grain cereals	water soluble
D	bones, teeth	fish oils, milk	fat soluble
E	not confirmed	liver, egg yolk	fat soluble
K	helps blood to clot	green, leafy vegetables	fat soluble
niacin	prevents pellagra	whole-grain cereals	water soluble
riboflavin	skin	whole-grain cereals	destroyed by alkali (bases)

Completion Questions

1. green plant	5. essential amino acids	9. vitamin K
2. proteins	6. vitamin C	10. proteins
3. carbohydrates	7. vitamin E	
4. amino acids	8. vitamin A	

Multiple-Choice

1. c
2. d
3. b

4. b
5. a
6. b

7. a
8. b
9. a

True or False

11-2b 1. carbohydrate
11-2i 2. T

11-2k 3. vitamin E
11-3a 4. proteins

Find the Facts

(The sections are indicated next to each answer above.)

Matching

3. nutrient
5. amino acids
1. fats
2. photosynthesis

6. proteins
7. vitamin C
4. vitamin D

Chapter 12 Exercises

Reading Power

1. b
2. c
3. d

4. a
5. c
6. d

7. a
8. d
9. a

Mind Expanders

Answers will vary.

Completion Questions

1. copolymer
2. rubber

3. the silicones
4. addition, condensation

5. condensation

True or False

Intro. 1. T
12-1   2. T
12-3d   3. T
12-2d   4. T
12-4   5. silicone

12-2d   6. T
12-5   7. incinerated
12-2d   8. elastomer
12-5   9. T
12-2e 10. T

Find the Facts

(The answers are listed with the True/False answers above.)

Multiple-Choice

1. c   2. b   3. a   4. a   5. d

Matching

3. thermoplastic
4. GR-S
1. nylon
2. PVC

Chapter 13 Exercises

Reading Power

1. b	4. b	7. d	10. d
2. c	5. c	8. a	
3. d	6. a	9. c	

Mind Expanders

Answers will vary.

Completion Questions

13-8	1.	lead	13-8	6.	sewage	
Intro.	2.	ecology	13-1	7.	dust	
13-1	3.	nitrogen, oxygen	13-7c, e	8.	carbon monoxide	
13-1	4.	aerosols	13-7b	9.	carbon dioxide	
13-1	5.	space	13-7e	10.	lead	

Find the Facts

(The supporting section numbers are indicated next to each answer above.)

Matching

3. pollution   5. greenhouse effect
1. ecology   2. aerosols
4. combustion

Multiple-Choice

1. c	4. d	7. b	10. c
2. b	5. b	8. d	
3. b	6. d	9. b	

True or False

1. old

2. automobile

3. T

4. T

5. carbon dioxide

Chapter 14 Exercises

Reading Power

1. b
2. c
3. a
4. d
5. c
6. d
7. c
8. a
9. b
10. a

Mind Expanders

Answers will vary.

Completion Questions

14-1 1. soap

14-9 2. oxidation

14-7 3. dry cleaning

14-5 4. lye

14-4 5. detergent

Multiple-Choice

14-4 1. c

14-4 2. a

14-3 3. c

14-3 4. d

14-9 5. d

True or False

Intro. 1. fragrant oils

14-9 2. sulfur dioxide

14-1 3. T

14-1 4. soap

14-2 5. T

Find the Facts

(The sections that support each answer are indicated next to each answer above.)

Matching

3. soap

5. emulsion

4. detergent

2. biodegradable

6. dry cleaning

7. bleaching

1. nascent oxygen

Periodic Table of the Elements

Atomic Number — 6 → Carbon
Element —
Symbol — **C**
Atomic Weight — 12.011
Electron Distribution — 2 4

1	2	3	4	5	6	7	8	9	10	11	12	13	14	15	16	17	18
Hydrogen 1 **H** 1.0079																	Helium 2 **He** 4.00260
Lithium 3 **Li** 6.941	Beryllium 4 **Be** 9.01218											Boron 5 **B** 10.81	Carbon 6 **C** 12.011	Nitrogen 7 **N** 14.0067	Oxygen 8 **O** 15.9994	Fluorine 9 **F** 18.998403	Neon 10 **Ne** 20.179
Sodium 11 **Na** 22.98977	Magnesium 12 **Mg** 24.305											Aluminum 13 **Al** 26.98154	Silicon 14 **Si** 28.0855	Phosphorus 15 **P** 30.97376	Sulfur 16 **S** 32.06	Chlorine 17 **Cl** 35.453	Argon 18 **Ar** 39.948
Potassium 19 **K** 39.0983	Calcium 20 **Ca** 40.08	Scandium 21 **Sc** 44.9559	Titanium 22 **Ti** 47.90	Vanadium 23 **V** 50.9415	Chromium 24 **Cr** 51.996	Manganese 25 **Mn** 54.9380	Iron 26 **Fe** 55.847	Cobalt 27 **Co** 58.9332	Nickel 28 **Ni** 58.70	Copper 29 **Cu** 63.546	Zinc 30 **Zn** 65.38	Gallium 31 **Ga** 69.72	Germanium 32 **Ge** 72.59	Arsenic 33 **As** 74.9216	Selenium 34 **Se** 78.96	Bromine 35 **Br** 79.904	Krypton 36 **Kr** 83.80
Rubidium 37 **Rb** 85.4678	Strontium 38 **Sr** 87.62	Yttrium 39 **Y** 88.9059	Zirconium 40 **Zr** 91.22	Niobium 41 **Nb** 92.9064	Molybdenum 42 **Mo** 95.94	Technetium 43 **Tc** (98)	Ruthenium 44 **Ru** 101.07	Rhodium 45 **Rh** 102.9055	Palladium 46 **Pd** 106.4	Silver 47 **Ag** 107.868	Cadmium 48 **Cd** 112.41	Indium 49 **In** 114.82	Tin 50 **Sn** 118.69	Antimony 51 **Sb** 121.75	Tellurium 52 **Te** 127.60	Iodine 53 **I** 126.9045	Xenon 54 **Xe** 131.30
Cesium 55 **Cs** 132.9054	Barium 56 **Ba** 137.33	Lanthanum 57 **La** ★ 138.9055	Hafnium 72 **Hf** 178.49	Tantalum 73 **Ta** 180.948	Tungsten 74 **W** 183.85	Rhenium 75 **Re** 186.207	Osmium 76 **Os** 190.2	Iridium 77 **Ir** 192.22	Platinum 78 **Pt** 195.09	Gold 79 **Au** 196.9665	Mercury 80 **Hg** 200.59	Thallium 81 **Tl** 204.37	Lead 82 **Pb** 207.2	Bismuth 83 **Bi** 208.9804	Polonium 84 **Po** (209)	Astatine 85 **At** (210)	Radon 86 **Rn** (222)
Francium 87 **Fr** (223)	Radium 88 **Ra** 226.0254	Actinium 89 **Ac** ▲ 227.0278	Unnilquadium 104 **Unq** (261)	Unnilpentium 105 **Unp** (262)	Unnilhexium 106 **Unh** (263)	Unnilseptium 107 **Uns** (267)	Unniloctium 108 **Uno** (265)	Unnilenium 109 **Une** (266?)									

Lanthanide series ★

Cerium 58 **Ce** 140.12	Praseo-dymium 59 **Pr** 140.9077	Neodymium 60 **Nd** 144.24	Promethium 61 **Pm** (145)	Samarium 62 **Sm** 150.36	Europium 63 **Eu** 151.96	Gadolinium 64 **Gd** 157.25	Terbium 65 **Tb** 158.9254	Dysprosium 66 **Dy** 162.50	Holmium 67 **Ho** 164.9304	Erbium 68 **Er** 167.26	Thulium 69 **Tm** 168.9342	Ytterbium 70 **Yb** 173.04	Lutetium 71 **Lu** 174.967

Actinide series ▲

Thorium 90 **Th** 232.0381	Protactinium 91 **Pa** 231.0359	Uranium 92 **U** 238.029	Neptunium 93 **Np** 237.0482	Plutonium 94 **Pu** (244)	Americium 95 **Am** (243)	Curium 96 **Cm** (247)	Berkelium 97 **Bk** (247)	Californium 98 **Cf** (251)	Einsteinium 99 **Es** (254)	Fermium 100 **Fm** (257)	Mendelevium 101 **Md** (258)	Nobelium 102 **No** (259)	Lawrencium 103 **Lr** (260)

THE CHEMICAL ELEMENTS

(Atomic masses in this table are based on the atomic mass of carbon-12 being exactly 12.)

NAME	SYMBOL	ATOMIC NUMBER	ATOMIC MASS†	NAME	SYMBOL	ATOMIC NUMBER	ATOMIC MASS†
Actinium	Ac	89	(227)	Neodymium	Nd	60	144.2
Aluminum	Al	13	27.0	Neon	Ne	10	20.2
Americium	Am	95	(243)	Neptunium	Np	93	(237)
Antimony	Sb	51	121.8	Nickel	Ni	28	58.7
Argon	Ar	18	39.9	Niobium	Nb	41	92.9
Arsenic	As	33	74.9	Nitrogen	N	7	14.01
Astatine	At	85	(210)	Nobelium	No	102	(255)
Barium	Ba	56	137.3	Osmium	Os	76	190.2
Berkelium	Bk	97	(247)	Oxygen	O	8	16.00
Beryllium	Be	4	9.01	Palladium	Pd	46	106.4
Bismuth	Bi	83	209.0	Phosphorus	P	15	31.0
Boron	B	5	10.8	Platinum	Pt	78	195.1
Bromine	Br	35	79.9	Plutonium	Pu	94	(244)
Cadmium	Cd	48	112.4	Polonium	Po	84	(210)
Calcium	Ca	20	40.1	Potassium	K	19	39.1
Californium	Cf	98	(251)	Praseodymium	Pr	59	140.9
Carbon	C	6	12.01	Promethium	Pm	61	(145)
Cerium	Ce	58	140.1	Protactinium	Pa	91	(231)
Cesium	Cs	55	132.9	Radium	Ra	88	(226)
Chlorine	Cl	17	35.5	Radon	Rn	86	(222)
Chromium	Cr	24	52.0	Rhenium	Re	75	186.2
Cobalt	Co	27	58.9	Rhodium	Rh	45	102.9
Copper	Cu	29	63.5	Rubidium	Rb	37	85.5
Curium	Cm	96	(247)	Ruthenium	Ru	44	101.1
Dysprosium	Dy	66	162.5	Samarium	Sm	62	150.4
Einsteinium	Es	99	(254)	Scandium	Sc	21	45.0
Erbium	Er	68	167.3	Selenium	Se	34	79.0
Europium	Eu	63	152.0	Silicon	Si	14	28.1
Fermium	Fm	100	(257)	Silver	Ag	47	107.9
Fluorine	F	9	19.0	Sodium	Na	11	23.0
Francium	Fr	87	(223)	Strontium	Sr	38	87.6
Gadolinium	Gd	64	157.2	Sulfur	S	16	32.1
Gallium	Ga	31	69.7	Tantalum	Ta	73	180.9
Germanium	Ge	32	72.6	Technetium	Tc	43	(97)
Gold	Au	79	197.0	Tellurium	Te	52	127.6
Hafnium	Hf	72	178.5	Terbium	Tb	65	158.9
Helium	He	2	4.00	Thallium	Tl	81	204.4
Holmium	Ho	67	164.9	Thorium	Th	90	232.0
Hydrogen	H	1	1.008	Thulium	Tm	69	168.9
Indium	In	49	114.8	Tin	Sn	50	118.7
Iodine	I	53	126.9	Titanium	Ti	22	47.9
Iridium	Ir	77	192.2	Tungsten	W	74	183.9
Iron	Fe	26	55.8	Unnilenium	Une	109	(266?)
Krypton	Kr	36	83.8	Unnilhexium	Unh	106	(263)
Lanthanum	La	57	138.9	Unniloctium	Uno	108	(265)
Lawrencium	Lr	103	(256)	Unnilpentium	Unp	105	(262)
Lead	Pb	82	207.2	Unnilquadium	Unq	104	(261)
Lithium	Li	3	6.94	Unnilseptium	Uns	107	(267)
Lutetium	Lu	71	175.0	Uranium	U	92	238.0
Magnesium	Mg	12	24.3	Vanadium	V	23	50.9
Manganese	Mn	25	54.9	Xenon	Xe	54	131.3
Mendelevium	Md	101	(258)	Ytterbium	Yb	70	173.0
Mercury	Hg	80	200.6	Yttrium	Y	39	88.9
Molybdenum	Mo	42	95.9	Zinc	Zn	30	65.4
				Zirconium	Zr	40	91.2

†Numbers in parentheses give the mass number of the most stable isotope.

Index

A

acetic acid, 130–32, 133
acetone, 253
acetylene, 238, 242–43
acidosis, 270
acid rain, 152, 166, 316–17, 320, 333
acids, 129–38, 143–45, 152, 181, 256–59
acrolein, 275–76, 280
actinides, 49
addition reactions, 297–98, 300, 303
aerosols, 313, 317, 334
agate, 182
air, 34, 313–18, 321, 332–33
air conditioners, 322
Alaska pipeline, 214
alcohols, 14, 97, 101, 104, 124–25, 156, 250–
    59, 276, 278, 291, 322–23, 346–47
aldehydes, 252, 256, 258–59
aliphatic hydrocarbons, 237–38, 242–45, 250
alkali metals, 49, 53
alkaline earth metals, 49, 53
alkalis, 132, 274, 343–44, 347
alkanes, 237–41, 243–45
alkenes, 237–39, 241–42, 244–45
alkynes, 237–38, 242, 244–45
alloys, 201, 211–12, 214–15, 221, 227
alum, 109
alumina, 225–26
aluminum, 74–75, 85, 155, 179, 182, 191, 202,
    225–28, 347
amide linkage, 300, 303
amino acids, 270–71, 279–80, 293, 303
ammonia, 32, 101, 133, 348, 352
ammonium, 73, 132, 135–36, 334
anaerobic bacteria, 319, 328
antifreeze, 104
antimony, 51
apatite, 272
aquifers, 325
argon, 51, 227, 313
aromatic hydrocarbons, 237–38, 244–45
Arrhenius, Svante, 130
arsenic, 51
ascorbic acid, 274

astatine, 50–51
atmospheric dust, 34–35, 312–15, 317, 332
atomic mass, 291
atomic number, 46–47, 48, 51, 52, 53, 55, 63,
    158, 166
atomic weight, 46–47, 48, 52, 53, 55, 56, 85–
    88, 151, 158
atoms, 43–44, 130, 271
    in carbon compounds, 245–46, 258
    combining of, 63–69, 71–76, 85–89, 151
    in metals, 202, 205, 223
    nature of, 44–47, 48–51, 52, 55, 214
    and nuclear radiation, 156, 159, 161,
        163–64, 166, 167, 184
    in polymers, 298, 300–01, 303
    in solutions, 95, 99, 102, 110
attractive force, 32–35, 103
automobiles, 104, 214, 294, 296, 318, 320–23,
    329, 333, 345
avidin, 275
Avogadro, Amedeo, 90
Avogadro number, 90–91

B

Bakelite, 293, 300, 303
Bakelite, Leo, 300
baking powder, 276, 278
bases, 132–38
batteries, 135–36, 186
bauxite, 225, 228
Becquerel, Henri, 156
Bell Telephone Company, 188
benzene, 238, 321, 348
beriberi, 272–73, 280
beryllium, 51
Bessemer process, 210–11, 215, 212
Bhopal accident, 323
bile juice, 96
biochemical oxygen-demand (BOD) test,
    329, 334
biochemistry, 237, 245

biodegradability, 302–03, 325, 330, 331, 333–34, 347, 352
biotin, 275
blast furnace, 186, 206, 208, 209, 214
bleaching agents, 349–52
blood, 134, 269–70, 272, 275, 281, 320
Bohr, Niels, 45, 47
boiling point, 14, 104, 110, 239–40
boron, 51, 58, 164
botulism, 279
breeder reactor, 163
brick, 180, 186, 189
brimstone, 293, 316
bromide/bromine, 50, 51, 321
bronze, 221, 228
Bronze Age, 202, 219
Brown, Basil, 273
burning—see combustion
butadiene, 300
butane/butene/butyne, 238, 239, 241, 243

C

calcination, 223, 224, 229
calcium, 109, 132, 181, 184, 188–91, 242, 272, 275, 346, 350–51
Caligula, 221
carbohydrates, 145, 146, 253, 258–59, 267–68, 278, 280, 281, 282, 287, 293
carbon, 48, 51, 58, 181, 191, 267–68, 271, 298, 303
    atomic structure of, 88-89, 161
    compounds (organic chemistry), 67, 235–45, 250–59, 321, 345, 348
    in metals, 209, 211–16, 223, 225–26, 228
carbon dioxide, 106, 181, 226, 253, 322
    in air, 104, 152, 166, 189, 191, 313, 315, 316, 324, 330–33
    in foods, 149–50, 267–68, 276, 278
carbon monoxide, 209, 211, 226, 314, 320–21, 322, 333
carbonic acid, 130, 136, 181, 315
carboxyl group, 256, 259
carcinogens, 159, 162, 322, 331, 333, 348
career information, 4, 23, 175, 232, 288, 339, 356
carotene, 273
carotenoids, 273
case hardening, 214, 216
catalytic converters, 321, 322
cells (of the body), 270–72, 275, 279, 313, 320

celluloid, 291, 302
cellulose, 265, 278, 291–92, 301
Celsius, Anders, 14
Celsius scale, 14–16
cement, 179, 182, 185–86, 188–89, 191
centigrade scale, 14–16
ceramics—see clay
Chadwick, James, 45
charge, 206, 208–09, 211–12, 216
chemical burns, 132, 137, 347
chemical change, 30–31, 36
chloride, 65–66, 102, 130, 135–36
chlorine, 50, 51, 102, 227, 302, 321, 329, 331, 349
    atomic structure of, 64–67, 69
chlorofluorocarbons, 317, 322
chlorophyll, 273
cholesterol, 270–71, 274
chromium, 186, 213
cigarette smoking, 320
clay, 179–80, 184–85, 189, 191, 225
cleansers, 346–47, 351–52
Cleopatra, 343-44
clouds, 34–35, 313, 315
coagulation, 275, 281
coal, 149, 151–52, 156, 165–66, 316–17, 322 330
cobalt, 186, 188
colloids, 99–100, 110–11
combustion, 186, 313–14, 318, 321–22, 326, 332, 348
composting, 329
compounds, acid, 130–32
    carbon, 235–42, 244–45, 252–58
    food, 267–68, 279
    metallic, 202, 205–06, 214, 225
    nature of, 63–65, 67–69, 72–76, 86–90, 109, 182, 191, 301
concrete, 184, 191
condensation, 34–36, 98, 110, 297, 300–01, 315
construction materials, 179–92
controlled experiments, 5–6, 16
cooking, 237, 274–81, 313–14, 279–81, 313–14
copolymers, 301, 303
copper, 155, 202–03, 205, 207, 221–22, 252, 272, 329
Copper Age, 202
corrosion, 214, 215, 216, 322
cosmic dust, 317
cosmic rays, 161

covalent bonds, 67–69, 73, 76–77, 236, 244
cryolite, 226, 227–28, 315
crystallization, 35, 71, 103, 106, 109, 111, 227–28, 315
cubit, 10–11
Curie, Marie and Pierre, 156
cyanide, 302

D

dacron, 301
Dalton atomic theory, 44, 49
Dalton, John, 44
DDT, 330–31, 333
decantation, 99, 110–11
decay bacteria—see microorganisms
Democritus, 43–44
detergents, 107–08, 325, 329, 333, 346–47, 351–52
diabetes, 270
diatomaceous earth (diatomite), 182, 191
diatoms, 182, 185
digestion, 100, 269–73, 275, 280
diseases, 6, 95, 152, 159–61, 270–71, 280, 329, 331, 343
dissolving, 103–11, 122, 124–25
distillation, 98, 110–11, 251
drain cleaners, 132, 347, 351
dry cleaning, 348, 352, 353
ductility, 209
dust—see atmospheric dust

E

ecology, 311–34
Egyptian civilization, 10, 179, 184, 188
Eijkman, Dr. Christiaan, 272
elasticity, 293, 302
elastomers, 292–93, 303–04
electric-furnace process, 211–13, 215
electricity, 70–71, 149, 152–53, 155, 163–64, 226, 323
electrolytic refining, 222–24, 228
electron-dot diagrams, 67–68, 74
electrons, 44–50, 51, 49–54, 63–66, 68–72, 74, 76, 102, 130, 156–59, 205, 214, 222, 236, 271
   valence electrons, 49–50, 55, 64–69, 72, 76–77, 202, 205, 236
elements, 43, 71, 75, 182–83, 191, 202, 235, 244, 267, 271, 298, 301, 313, 316, 332

elements *(cont.)*
   atomic structure of, 44, 46–55, 63–64, 66–67, 73–74, 76, 85, 89, 132, 158–59, 161, 164
   as metals, 202, 211, 215, 220–21, 227
Elizabeth I, Queen of England, 343
empirical formula, 88–89
emulsions, 100, 345–46, 348, 351–52
energy, 27, 32–34, 36, 102–03, 227, 293
   in food cycle, 267–69, 272, 279–81
   sources of, 98, 149–56, 161, 165–70, 278–79
energy levels (orbits), 45, 47, 53
English system, 10, 15
environment—see ecology
ergosterol, 274
esters, 257, 259
ethanol, 256, 259, 322-23
ethene, 238, 241–43, 256
ethylene, 104, 298
ethyne, 242–43
eutrophication, 329, 333–34
evaporation, 34–36, 315
explosion, 314, 332, 348

F

Fahrenheit, Gabriel Daniel, 14
Fahrenheit scale, 14
Faraday, Michael, 293
fats, 100, 103, 268–69, 271–75, 279–81, 343–45, 351
fatty acids, 272, 275, 345–46
feldspar, 186, 191
fermentation, 156, 253–54, 258–59, 276–77, 323, 343
fibers, 187–88, 293–94, 297, 303
filtration, 99, 110–11, 325
Firestone Tire Company, 293
fission—see nuclear reactions
flotation process, 206–07, 209, 215–16, 222–23
fluorine, 50, 51, 271
flux, 209, 216
fly ash, 321, 331, 334
fog, 315
food, chemistry of, 134, 138, 149, 161, 166, 257, 267–81
food chain, 325, 329–30, 333–34
food supply, 95, 267, 279
forest fires, 313, 315, 317

formaldehyde, 252, 300. 322
formulas, 71–76, 85–86, 88–89, 129, 132, 203, 239, 243–45, 252, 259
formula weight, 85, 123
freezing point, 14, 33, 36, 104, 110, 239
fuels, 163–64, 165–67, 237, 313, 315, 317–18, 322, 331–33
    fossil fuels, 149, 151–52, 156, 165–67
Funk, Casimir, 272
fusion—see nuclear reactions

G

galena, 222, 228
gangue, 206, 209, 215, 216
gas, natural—see methane
gasohol, 323
gasoline, 222, 313, 321–23
gases, in air, 313, 317–18, 331
    in carbon compounds, 341–44
    poisonous, 152, 302–03, 314–18, 320–21, 333
    properties and states of, 29, 32, 34–36, 50–51, 324
    in solutions, 95, 97, 104, 106, 108–11
gels, 100
geothermal energy, 155–56, 167
germanium, 51
germ theory of disease, 6
germs, 6, 95
glass, 67, 179, 188, 191–92
glucose, 269–70, 280, 292
glycerol, 272, 275–76. 343–45, 351
gold, 202, 205, 223, 225
Goldberger, Dr. Joseph, 274
Goodyear, Charles, 293, 294
granite, 179
graphite, 164, 226
Greek civilization, 313
greenhouse effect, 152, 311, 317–18, 321, 322, 333
GR-S, 300
gypsum, 190–91

H

half-life, 161–62, 167
Hall, Charles Martin, 225–26
Hall, process, 226, 228
halogens, 50–51
helium, 46, 49, 51, 151, 156, 164, 227
hematite, 206

hexane, 243
Hooker Chemical Company, 331
Hyatt, John Wesley, 291
hydrocarbons, 237–41, 244–45, 250–58, 313, 321, 322
hydrochloric acid, 275
hydrogen, 151, 165, 271, 298
    atomic structure of, 43, 46, 49, 63, 67, 69, 88–89, 130, 138, 164–65, 205, 220–21, 223
    in carbon compounds, 237–45, 251–52, 256–58
    compounds, 69, 130–31, 135–36, 302, 331, 350–51
    properties of, 324
    in water, 64, 71, 74, 86–87, 142–44, 269, 315, 347
hydrogen, liquid, 323
hydronium ions, 130, 133, 135, 137–38
hydroxide ions, 132, 135, 137, 142, 144
hydroxyl groups, 250
hypo, 106
hypothesis (defined), 5–6, 16

I

ilmenite, 227
incineration, 302–03, 331–33
indicators, 132–33, 137–38
industrial wastes, 321, 325–26, 329–30, 331–33
inert elements, 51–52
insecticides, 323, 325, 333
iodine, 50–51, 161
ionization detectors, 159
ionization equation, 204
ions, 65–66, 69, 71, 73, 76–77, 92, 159, 164, 256, 346, 351
    in acids and bases, 130–33, 135, 137–38, 142–44
    in metals, 204–05, 223, 225–26, 228
    in solutions, 95, 99, 102–04, 110, 123
iron, 33, 186, 188, 202–03, 214, 215–19, 313, 346, 351
    extracting, 206, 208–09, 211–12, 214–15, 327
    as a nutrient, 272, 329
Iron Age, 202
iron oxide, 206, 214
Isabella, Queen of Spain, 343
isobutane, 239
isomers, 239–40, 244–45, 251
isopropanol, 251

isotopes, 47, 52–53, 151, 158–63
ivory, 291

J

Javelle water, 349

K

kaolin, 186
ketones, 252–53, 258–59
kinetic molecular theory, 32–33, 36
Kroll process, 227
krypton, 51

L

lab reports, 7–9, 15
lab safety rules, 6–7, 15
lactose, 134
landfills—see sanitary landfills
lanthanides, 49
laundering, 107–09, 206, 345–47, 349, 351
Lavoisier, Antoine, 313
law (defined), 5
law of definite proportions, 86–87
lead, 156, 158, 161, 202, 220–23, 228, 322,
    325, 329, 333
lecithin, 272
lightning, 316
lime, slaked, 132, 189
limestone, 179, 181–82, 184, 188–89, 191–92,
    209, 212
Lind, Dr. James, 274
liquids, properties and states of, 29, 32–37,
    50–52, 239–40
    in solutions, 95, 97, 103, 104, 107, 110
lithium, 51, 132, 137, 345
litmus, 132–33, 135–37
Love Canal, 331
lucite, 292
lye, 132, 343–44, 347–48

M

macromolecules, 291, 304
magnesium, 132, 214, 346, 351
magnetite, 206
magnetron, 278–79

malachite, 223–24
malleability, 209
mammoths, 302
manganese, 211, 329
marble, 179, 192
mass number, 47, 50
matte, 223–24, 228–29
matter, defined, 27–29, 31, 36
    states of, 29, 32–36, 240
Mayan Indians, 293
measuring systems, 10–15
melting point, 33–34, 36–37, 240
Mendeleev, Dmitri Ivanovich, 48
mercury, 13–14, 329, 333
metallic activity, 202, 204–05, 215–16, 225
metalloids, 50, 51, 53
metallurgy, 201, 206, 215, 232
metals, 49–51, 53, 64–66, 69–73, 75–77, 95,
    132, 135, 156, 166, 182, 186, 191, 201–15,
    221–29, 278–79, 313, 329–30, 333
methaldehyde, 252
methane (natural gas), 101, 156, 165, 237–
    38, 240, 244, 313–14, 317, 331, 333
methanol, 251–52, 322–23
methyl isocyanate, 323
metric system, 9–10, 12–16
    conversion table, 15
microoganisms, 134, 279–81, 302, 314, 325,
    329, 331, 333–34, 347
microwaves, 278–79, 281
milk of magnesia, 132
minerals, 161, 182, 186, 190–92, 202–03, 209,
    214, 222–23, 225, 228, 317
    as nutrients, 268, 272, 275, 279–82
mining, 151–52, 166, 206–07, 209, 215, 316,
    327, 329–30, 333
mole and molarity, 90–91, 123–26, 141
molecular weight, 86–87, 90, 123–26, 240,
    291
molecules, in acids and bases, 130–32, 142
    in carbon compounds, 150, 235, 239,
        241–44, 256–57
    in foods, 267–68, 276
    motion of, 31–37
    nature of, 43–44, 52, 63–64, 66, 71, 68–69,
        78, 90–91, 159, 227, 291, 304
    in polymers, 291–92, 300, 302
    in soaps and detergents, 345, 347
    in solutions, 95, 99–104, 107, 110, 112,
        123–24
molybdenum, 203

monomers, 291–92, 297–304
mortar, 189, 191–92
mothballs, 34
Mt. St. Helens, 312

N

Napoleon I, 10
nascent (atomic) oxygen, 350, 352
negative charge, 44–45, 52, 65–66, 71, 73, 102
  156, 158, 205, 222, 225, 346
neon, 51, 313
Nero, 221
neutralization, 134–35, 138
neutrons, 45–47, 52–53, 156, 161, 163–65
niacin, 274
nitrates, 316
nitric acid, 130–31, 137, 316, 322
nitrogen, 51, 67, 102, 104, 214, 300, 313, 317,
  321, 322, 332–33
nonmetals, 50, 51, 53, 68–72, 72–73, 75–77,
  102, 135, 205
nonpolar bonds, 67, 76–77
nonpolar molecules, 101–03, 110, 345
nuclear medicine, 160–61
nuclear reactions, 151–52, 157–67, 184, 188
  (see also radioactivity)
nuclei, atomic, 45–47, 49–50, 51–52, 53–54,
  63–65, 151, 156–58, 161, 163–65
nutrients and nutrition, 267–81
nylon, 291–92, 294, 300–01, 303

O

Oersted, Hans Christian, 225
oils, 100, 103, 110, 271–72, 275–76, 280–81, 302,
  315, 321, 322, 325, 333, 344–48, 351–52
opal, 182
open-hearth process, 211–12, 215
ores, 163, 165–66, 203, 206–07, 209, 215, 216,
  221–22, 225, 228, 321, 327
organic acids—see acids
organic chemistry, 235–36, 244–45, 252, 258
orlon, 302
osmosis, 325
oven cleaners, 348, 351
ovomucoid, 275
oxidation, 214, 250–52, 256–59, 271, 275,
  313–14, 316, 349–50, 352
oxides, 49, 53, 164, 186, 188, 191, 206, 211, 222–
  26, 228, 252, 316, 321, 322, 333, 343

oxygen, 102, 182, 191, 226–27. 269, 271, 301,
  303–04, 320
  in air, 104, 214, 216–17, 313–16, 332
  atomic structure of, 43, 50, 51, 52, 63, 67,
    88–89, 350, 352
  in hydrocarbons, 250–53, 256–58
  kinetic energy of, 33
  properties of, 324
  in water, 64, 71, 74, 86–87, 325, 329–30,
    333, 351
ozone, 313, 317, 325a

P

Pasteur, Louis, 5–6
Pauling, Dr. Linus, 274
pectin, 278
pellagra, 274, 280
pentane, 238, 240
pentanol, 258
periodic table, 48–50, 51, 52–53, 381
pesticides, 330–31, 333
pH, 133–34, 137–38, 142–43
phenol, 300
pheromones, 330, 334
phlogiston, 313
phosphates, 325, 329, 333
phosphorus, 51, 209, 223, 272
photosynthesis, 149–50, 166, 267–68, 280,
  281
physical change, 29–31, 33–36
plasma, 164
plaster of paris, 184, 190–92
plastics, 291–94, 297–305, 330–31
plutonium, 163–64
poisons, 95, 98, 152, 163, 221–22, 228, 251–
  52, 256, 259, 302–03, 315–16, 321–22, 325,
  329, 330–32, 333, 350, 352
polar bonds, 66, 68–69, 77
polar molecules, 101–04, 110, 112, 130, 142,
  253, 345
pollution, 95, 152, 155, 166, 227, 302,
  315–23, 325, 329–334
polonium, 50
polyamides, 300
polychlorinated biphenyls (PCBs), 331–32
polyester, 257, 291, 299
polyethylene, 291, 298, 303
polymers, 291–93, 297–304
polystyrene, 291–92
polyvinylchloride (PVC), 292, 331

positive charge, 45, 52, 54, 65–66, 71, 73, 102, 156, 158, 205, 223, 225–26, 346
potassium, 132, 137, 188, 191, 225, 343
preserving foods, 278–80
propane/propene/propyne, 238, 241, 243, 251
propanol, 251, 253
protactinium, 159
proteins, 268, 270–71, 275, 280–81, 300, 304, 316, 320, 347
proton number—see atomic number
protons, 45–47, 52, 53, 54, 63, 65, 130, 156, 158, 164–65
public health, 274, 343

Q

quartz, 182
quenching, 213, 216

R

radicals, 73, 75–77, 142, 251–52, 257, 302
radioactivity (radiation), 50, 156–67, 325
radium, 156
radon, 51
rain, 34–35, 152, 165–66, 313, 315–17, 320, 331, 333
rancidity, 271
reactions, 76, 208, 250, 252, 259, 297–303, 314
reclamation, 330
recycling, 267, 330
reduction, 206, 216, 257, 259, 349–352
rem, 160
retinoic acid, 273
reverberatory furnace, 222, 229
riboflavin, 274
rickets, 274
roasting, 222–23, 229
Rocky Mountain Arsenal, 325
Roman civilization, 180, 221–22, 325, 333, 343
rubber, 291, 293–96, 300, 302–04
rusting, 214, 215
Rutherford, Ernest, 45, 156
rutile, 227

S

safety rules for labs, 6–7, 15
saltation, 317, 334

salts, 14, 30, 65–66, 71, 92, 95, 102, 104, 109, 135–36, 138, 156, 164, 182, 188, 276, 278–79, 317, 346
salt water, 214, 325
sand, 67, 182, 184, 188–89, 191, 209, 212
sandstone, 192
sanitary landfills, 237, 303, 325, 328, 330, 331, 333
saponification, 343, 348, 351
saran, 292
scientific method, 3, 5, 15–16
scientific notation, 91–92
scintillation counter, 159
scurvy, 272, 274
selenium, 50, 51, 188
sewage treatment, 325, 327–30, 333, 347
shells—see energy levels (orbits)
silica (silicon dioxide), 182, 192
silicon, 51, 67, 182, 191, 209, 301, 303–04
silicone, 301–04
silver, 202, 205, 223
slag, 186, 192, 208–09, 212, 216
snow, 14, 29, 35, 313, 315
soap, 100, 107, 343–47, 351–52
sodium, 132
    atomic structure of, 46, 64–66
    in compounds, 85, 102, 106, 132, 135–36, 159, 188, 191, 274, 343, 347
    in household chemicals, 343, 345–47, 349–50
sol, 100
solar energy, 98, 149–55, 161, 164, 165–67, 273, 311, 317
solid waste disposal, 331, 333
solids, properties and states of, 29, 32–37, 50–51, 239–40
    in solutions, 95, 97, 105, 107–09, 111
solubility product, 143
solutes, 95, 97–98, 103–07, 110–12, 122–25
solutions, 95–112, 122–25, 130, 133, 135, 144, 181, 189, 223, 315, 348–50
solvents, 95, 97–98, 103, 105–07, 110–12, 294, 348, 352
space shuttle, *Challenger*, 175–76
span, 10
spiegeleisen, 211
spontaneous combustion, 313–14, 332
Stahl, Georg Ernst, 313
stain and spot removal, 348–49, 352
starch, 269–70, 276, 278–80
steel, 164, 186, 209–10, 211–17, 214–17, 294

Stone Age, 222
styrene, 300
sublimation, 34–36, 227–28, 315
sucrose, 124–25, 270
sugar, 30–31, 101–03, 122–24, 253
    as a nutrient, 269–70, 276, 278–81
sulfides, 222–23, 228, 321, 329
sulfur, 29, 50, 51, 209, 222–23, 293, 316
sulfur dioxide, 152, 222, 316, 321, 324, 350–51
sulfuric acid, 130–31, 137, 152, 166, 257, 259,
    316, 321, 346–47
sunlight, 98–99, 159, 267–68, 273–74, 280,
    281, 311, 317–18 (see also solar energy)
suspensions, 99–100, 110, 112, 122–24. 315,
    345–46, 351

T

taconite, 206, 215
Takaki, Dr., 272
technetium, 160
teflon, 298
tellurium, 50, 51
temperature, heat treatments, 184, 186,
    213–16, 222–23, 228
    ignition temperature, 313–15, 332
    measuring, 13–16
    and molecular motion, 32–34, 36–37, 70,
        103–11, 164, 239, 302–03
tempering, 213, 216–17
thallium, 159, 161
theory (defined), 5–6, 16, 32
thermometers, 13–14, 32
thiamine, 273
Thompson, John J., 44, 156
tin, 202, 222
tincture, 97
tissue (of the body), 270–72, 280–81, 320,
    330, 350
titanium, 201–02, 227–28
Tokomak reactor, 164–65
transmutation reactions, 158, 166–67
transuranium elements, 52, 54
tritium, 151
trypsin, 275
tungsten, 207
Tyndall effect and test, 99–100

U

ultraviolet light, 302

uranium, 151, 156, 161–64, 203
urea, 236–37

V

valence electrons—see electrons
valences, 72–75, 77
    table of, 73
vanadium, 212
vapor—see gases
vaporization, 98, 110
vinyl cyanides, 302
vitamins, 268, 272–75, 280–81
volcanoes, 312–13, 316–17
vulcanization, 293, 304

W

water, in construction materials, 184, 186,
    189–90
    in environment, 95–98, 110, 315, 317, 332
    in foods, 149–50, 267–70, 272–74, 278–80
    freezing and boiling points, 14, 33
    heavy, 160
    kinetic energy of, 33
    molecular structure of, 43, 63–64, 67, 71, 74,
        86–87, 90–91, 130, 132, 135–38, 142–43
    pollution of, 221, 325, 329–33
    in solutions, 30, 99–110, 181, 223, 252–53,
        256–57, 300
    vapor—see rain
    washing and cleaning with, 107–09, 345–52
winemaking, 5–6, 99, 253–55, 278
Wøhler, Friedrich, 236–37
wood alcohol—see methanol

X

xenon, 51
xeropthalmia, 273
X rays, 156-57, 159

Y

yeast, 276, 278

Z

zinc, 135–36, 205, 214, 223, 329

Notes